Springer Monographs in Mathematics

Jeffrey Groah
Joel Smoller
Blake Temple

Shock Wave Interactions in General Relativity

A Locally Inertial Glimm Scheme for
Spherically Symmetric Spacetimes

 Springer

Jeffrey Groah
Department of Mathematics
Montgomery College
Conroe, TX 77384
USA

Joel Smoller
Department of Mathematics
University of Michigan
Ann Arbor, MI 48109
USA

Blake Temple
Institute of Theoretical Dynamics
University of California, Davis
Davis, CA 95616
USA

Mathematics Subject Classification (2000): 35L65, 35L67, 83C05

ISBN-13: 978-1-4419-2246-5   e-ISBN-13: 978-0-387-44602-8

Printed on acid-free paper.

9 8 7 6 5 4 3 2 1

springer.com

Preface

General relativity is the modern theory of the gravitational field. It is a deep subject that couples fluid dynamics to the geometry of spacetime through the Einstein equations. The subject has seen a resurgence of interest recently, partly because of the spectacular satellite data that continues to shed new light on the nature of the universe...Einstein's theory of gravity is still the basic theory we have to describe the expanding universe of galaxies. But the Einstein equations are of great physical, mathematical and intellectual interest in their own right. They are the granddaddy of all modern field equations, being the first to describe a field by curvature, an idea that has impacted all of physics, and that revolutionized the modern theory of elementary particles. In these notes we describe a mathematical theory of shock wave propagation in general relativity. Shock waves are strong fronts that propagate in fluids, and across which there is a rapid change in density, pressure and velocity, and they can be described mathematically by discontinuities across which mass, momentum and energy are conserved. In general relativity, shock waves carry with them a discontinuity in spacetime curvature. The main object of these notes is to introduce and analyze a practical method for numerically computing shock waves in spherically symmetric spacetimes. The method is *locally inertial* in the sense that the curvature is set equal to zero in each local grid cell. Although it formally appears that the method introduces singularities at shocks, the arguments demonstrate that this is not the case.

The third author would like to dedicate these notes to his father, Paul Blake Temple, who piqued the author's interest in Einstein's theory when he was a young boy, and whose interest and encouragement has been an inspiration throughout his adult life.

Blake Temple
Davis California
July 15, 2006

Contents

Shock Wave Interactions
in General Relativity

1

Introduction

These notes present a self contained mathematical treatment of the initial value problem for shock wave solutions of the Einstein equations in General Relativity. The first two chapters provide background for the introduction of a locally inertial Glimm Scheme in Chapter 3, a non-dissipative numerical scheme for approximating shock wave solutions of the Einstein equations in spherically symmetric spacetimes. In Chapter 4 a careful analysis of this scheme provides a proof of the existence of (shock wave) solutions of the *spherically symmetric* Einstein equations for a perfect fluid, starting from initial density and velocity profiles that are only locally of bounded total variation. To keep the analysis as simple as possible, we assume throughout that the equation of state is of the form $p = \sigma^2 \rho$, $\sigma = const$. For these solutions, the components of the gravitational metric tensor are only *Lipschitz continuous* functions of the spacetime coordinates at shock waves, and so it follows that these solutions satisfy the Einstein equations, as well as the relativistic compressible Euler equations, only in the weak sense of the theory of distributions. The existence theory presented here establishes the consistency of the initial value problem for the Einstein equations at the weaker level of shock waves, for spherically symmetry spacetimes.

The material of Chapter 4 is taken from the work of Groah and Temple [13], and relies on the results of Chapters 2 and 3. The material of Chapter 3 is taken from the work of Groah and Temple, [12], and Chapter 2 is taken from the work of Smoller and Temple [27]. The introductory material in Sections 1.1 and 1.2 of Chapter 1 is taken mostly from [30, 31], while the material in Sections 1.3 and 1.4 is from [12, 13].

Chapter 2 outlines the simplest possible setting for shock wave propagation in Special Relativity; namely, the case of a perfect fluid, under the assumption that the equation of state is given by $p = \sigma^2 \rho$, where the sound speed σ is assumed to be constant, $0 < \sigma < c$. The assumption that $\sigma < c$ ensures that wave speeds are uniformly bounded away from the speed of light for arbitrarily large densities, and σ bounded away from zero prevents the formation of vacuum states $\rho = 0$, a well known singularity in the compressible Euler equations, [26]. Although this is a model problem, it has a number of interesting applications in General Relativity. The case $\sigma^2 = c^2/3$ is the equation of state for the extreme relativisitic limit of free particles, as well as for pure radiation, and for radiation in thermal equilibrium with matter when the energy density of radiation dominates. In particular, this equation of state applies in the early universe, and has been derived as a model for the equation of state in a dense Neutron star, [35]. The equation of state $p = \sigma^2 \rho$ can also be taken to be the equation of state for isothermal flow, applicable to simple models of star formation, [29].

The main part of Chapter 2 is devoted to a proof of global existence of shock wave solutions of the initial value problem for the relativistic compressible Euler equations in *flat*, $(1 + 1)$-dimensional Minkowski spacetime, assuming σ is constant. The analysis establishes a special property of solutions of the relativistic compressible Euler equations when $\sigma = const$. Namely, in the absence of gravity, the total variation of $\ln \rho$ is non-increasing in time on weak solutions of the initial value problem. This property of weak solutions when $p = \sigma^2 \rho$ was first discovered in the non-relativistic regime by Nishida, [22], and extended to the relativistic case in [28]. (See also [19, 23, 11, 20].) The method used to obtain these results is based on the celebrated Glimm difference scheme, or random choice method. This Glimm scheme is an approximation method by which the solution is discretized into piecewise constant states, and approximated locally by solving the so called *Riemann problem* posed at the discontinuities. The novelty of this method is that new constant states are chosen by a random choice selection process. An advantage of this random choice process is that it eliminates dissipation effects which arise in classical finite difference schemes, (very difficult to analyze), and we prefer the random choice method over the front tracking method as the latter introduces errors that obscure the locally inertial character of the scheme when extended to general relativity. The main result on the solution of the Riemann problem is given in Theorem 4, and the general existence theorem based on the Glimm scheme is stated in Theorem 5.

In Chapter 3 we show that the spherically symmetric Einstein equations in standard Schwarzschild coordinates are weakly equivalent to a system of conservation laws with source terms. The main result is given in Theorem 9 which provides a locally inertial expression of the Einstein equations that admits a valid weak formulation. Interestingly, the Einstein equations themselves lack a differential equation for the timelike metric component, and the locally inertial formulation provides conserved quantities in terms of which the equations close. This is demonstrated in Theorem 8. (The results of Chapter 3 require no restriction on the equation of state.)

The locally inertial expression of the equations is amenable to study by the locally inertial Glimm scheme, and this is the subject of Chapter 4. In their locally inertial form, the conserved quantities are taken to be the flat Minkowski spacetime energy and momentum densities. Thus, in Chapter 4, we can exploit the estimates of Chapter 2 in the conservation law step (based on

the Riemann problem) of a fractional step Glimm scheme that is introduced
for the analysis of the initial value problem for the full (spherically symmetric)
Einstein equations, c.f. [19, 10, 26]. We prove in Chapter 4 that this fractional
step Glimm appoximation scheme converges to a weak solution of the Einstein
equations as the mesh length tends to zero. The main result on the existence
and regularity of large BV solutions constructed by the Glimm scheme is
stated in Theorem 11.

In these notes we interpret our fractional step method as a *locally inertial
Glimm scheme*, in the sense that it exploits the locally flat character of spacetime. That is, the scheme has the property that it solves the compressible
Euler equations of Minkowski spacetime *exactly* in locally inertial coordinate
frames, (grid rectangles), and the transformations between neighboring coordinate frames are accounted for by discontinuities at the coordinate boundaries.
In Chapter 1 we introduce General Relativity and the Einstein equations as a
locally flat theory, with an eye toward interpreting the numerical method of
Chapter 4 in these terms. Included in Chapter 1 is an introduction to shock
waves and perfect fluids. For a survey of existence theories for classical solutions of the Einstein equations, we refer the reader to the article [25]. For
a nice discussion of relativistic fluids we refer the reader to [1]. For other
references see [2, 3, 4, 5, 6, 16].

1.1 Introduction to Differential Geometry and General Relativity

In Einstein's theory of General Relativity, all properties of the gravitational
field are determined by the *gravitational metric tensor g*, a Lorentzian metric
that describes a continuous field of symmetric bilinear forms of signature
$(-1, 1, 1, 1)$, defined at each point of a four dimensional manifold M called
"spacetime." Freefall paths through the gravitational field are the geodesics of
the metric; the non-rotating vectors carried by an observer in freefall are those
vectors that are parallel transported by the (unique symmetric) connection
determined by g; spatial lengths of objects correspond to the lengths of the
spacelike curves that define their shape—length measured by the metric g; and
time changes for an observer are determined by the length of the observer's
timelike curve through spacetime, as measured by the metric g.

The length of a curve in spacetime is computed by integrating the element of arclength ds along the curve, where, in a given coordinate system on
spacetime, ds is defined by

$$ds^2 = g_{ij}dx^i dx^j. \tag{1.1.1}$$

Here we adopt the Einstein summation convention whereby repeated up-down
indices are assumed to be summed from 0 to 3. A coordinate system on spacetime is a regular map that takes a neighborhood U_x of spacetime to $\mathbf{R}^4$, x :

$U_x \to \mathbf{R}^4$. Since spacetime is a manifold, it can be covered by coordinate charts. We let $x = (x^0, x^1, x^2, x^3)$ denote both the coordinate map and the coordinates of a point $x(P) \in \mathbf{R}^4$. The functions $g_{ij}(x)$, $i, j = 0, 1, 2, 3$, are the x-components of the metric g. At each point x, the matrix g_{ij} determines the lengths of tangent vectors in terms of their components relative to the x-coordinate basis $\left\{ \frac{\partial}{\partial x^i} \right\}$. That is, in x-coordinates, the tangent vector to a curve $x(\xi)$, (as parameterized in x-coordinates), is given by $X(\xi) = \dot{x}^i \frac{\partial}{\partial x^i}$, so that along the curve $x(\xi)$, the increment dx^i in the x^i-coordinate, in the direction of the curve, is given by $dx^i = \dot{x}^i$. Thus, according to (1.1.1), the increment in arclength along a curve $x(\xi)$ is given in terms of the increment in the parameter ξ by

$$ds^2 = g_{ij} \dot{x}^i \dot{x}^j d\xi^2 = \|X(\xi)\|^2 d\xi^2,$$

so that, the length of an arbitrary vector $X = X^i \frac{\partial}{\partial x^i}$ is given by

$$\|X\|^2 = g_{ij} X^i X^j,$$

where again we assume summation over repeated up-down indices. We conclude that the length of a curve is just the integral of the g-length of its tangent vector along the curve. Under change of coordinates $x \to y$, a vector $X^i \frac{\partial}{\partial x^i}$ transforms to $X^\alpha \frac{\partial}{\partial y^\alpha}$ according to the tensor transformation laws

$$X^\alpha = \frac{\partial y^\alpha}{\partial x^i} X^i, \qquad \frac{\partial}{\partial y^\alpha} = \frac{\partial x^i}{\partial y^\alpha} \frac{\partial}{\partial x^i}. \tag{1.1.2}$$

(Our slightly ambiguous notation is that indices $i, j, k, \ldots$ label components in x-coordinates, and $\alpha, \beta, \gamma, \ldots$ label components in y-coordinates. So, for example, X^i is the x^i-component of the tangent vector X, X^α is the y^α-component of X, etc. This works quite well, but tensors must be re-labeled when indices are evaluated.) It follows that the metric tensor transforms according to the tensor transformation law

$$g_{\alpha\beta} = g_{ij} \frac{\partial x^i}{\partial y^\alpha} \frac{\partial x^j}{\partial y^\beta}. \tag{1.1.3}$$

That is, at each point, g transforms by the matrix transformation law

$$\bar{g} = A^t g A$$

for a bilinear form, because the matrix $A = \frac{\partial x^j}{\partial y^\beta}$ transforms the vector components of the y-basis $\left\{ \frac{\partial}{\partial y^\alpha} \right\}$ over to their components relative to the x-basis $\left\{ \frac{\partial}{\partial x^i} \right\}$. The Einstein summation convention keeps track of the coordinate transformation laws as in (1.1.2) and (1.1.3) so long as we keep the indices on coordinate functions "up" (as in x^i), coordinate basis indices "down" (as in $\frac{\partial}{\partial x^i}$), indices on vector components "up", (as in X^i so that $X = X^i \frac{\partial}{\partial x^i}$), indices on basis 1-forms "up" (as in dx^i), and indices on components of 1-forms down (as in ω_i so that $\omega = \omega_i dx^i$). In general, a tensor of type (k, l) is said to have k-contravariant indices (up) and l-covariant indices (down) if the components in a given coordinate system transform according to the tensor transformation law

$$T^{\alpha_1,\ldots,\alpha_k}_{\beta_1,\ldots,\beta_l} = T^{i_1,\ldots,i_k}_{j_1,\ldots,j_l} \frac{\partial y^{\alpha_1}}{\partial x^{i_1}} \cdots \frac{\partial y^{\alpha_k}}{\partial x^{i_k}} \frac{\partial x^{j_1}}{\partial y^{\beta_1}} \cdots \frac{\partial x^{j_l}}{\partial y^{\beta_l}}.$$

Here the (matrix) Jacobian satisfies $\frac{\partial x}{\partial y} = \left(\frac{\partial y}{\partial x}\right)^{-1}$, and by letting

$$g^{ij} = g_{ij}^{-1},$$

we can raise or lower an index by contracting the index with the metric; that is, for example,

$$T^i_j = T_{\sigma j} g^{\sigma i}$$

raises the index i. In the modern theory of differential geometry, $T^{i_1,\ldots,i_k}_{j_1,\ldots,j_l}$ are viewed as the components of the tensor products

$$\left\{ \frac{\partial}{\partial x^{i_1}} \otimes \cdots \otimes \frac{\partial}{\partial x^{i_k}} \otimes dx^{j_1} \otimes \cdots \otimes dx^{j_l} \right\},$$

which form a basis for the set of operators that act linearly on k copies of T^*M and l copies of TM, c.f. [8].

Freefall paths through a gravitational field are geodesics of the spacetime metric g. For example, the planets follow geodesics of the gravitational metric generated by the Sun, (approximated by the Schwarzschild metric beyond the surface of the Sun, and by the Tolman-Oppenheimer-Volkoff (TOV) metric inside the surface of the Sun, [33, 24]), and according to the standard theory of cosmology, the galaxies follow geodesics of the Friedmann-Robertson-Walker (FRW) metric. In spherical coordinates $x = (t, r, \theta, \phi)$, the Schwarzschild line element is given by

$$ds^2 = -\left(1 - \frac{2GM_0}{r}\right) dt^2 + \left(1 - \frac{2GM_0}{r}\right)^{-1} dr^2 + r^2 d\Omega^2, \qquad (1.1.4)$$

the TOV line element is given by

$$ds^2 = -B(r)dt^2 + \left(1 - \frac{2GM(r)}{r}\right)^{-1} dr^2 + r^2 d\Omega^2, \qquad (1.1.5)$$

and the FRW line element is given by

$$ds^2 = -dt^2 + R(t)^2 \left(\frac{dr^2}{1 - kr^2} + r^2 d\Omega^2\right). \qquad (1.1.6)$$

The line element determines the metric components g_{ij} through the identity (1.1.1). Here $\mathcal{G}$ denotes Newton's gravitational constant, M_0 denotes the mass of the Sun (or a star), $M(r)$ denotes the total mass inside radius r, (a function that tends smoothly to M_0 at the star surface), $B(r)$ is a function that tends smoothly to $1 - 2\mathcal{G}M_0/r$ at the star surface, $H = \frac{\dot{R}(t)}{R(t)}$ is the Hubble "constant", and $d\Omega^2 = d\theta^2 + sin^2(\theta)d\phi^2$ denotes the standard line element on the unit 2-sphere. (Here $2GM = \frac{2\mathcal{G}M}{c^2}$, and we take $c = 1$, [8].)

Each of the metrics (1.1.4)-(1.1.6) is a special case of a general spherically symmetric spacetime metric of the form

$$ds^2 = -A(r,t)dt^2 + B(r,t)dr^2 + 2D(r,t)drdt + C(r,t)d\Omega^2, \quad (1.1.7)$$

where A, B, C, D are arbitrary, smooth, positive functions. A spherically symmetric metric is said to be in standard Schwarzschild coordinates, (or the standard coordinate guage), if it takes the simpler form

$$ds^2 = -A(r,t)dt^2 + B(r,t)dr^2 + r^2 d\Omega^2. \quad (1.1.8)$$

It is well known that, (under generic conditions), there always exists a coordinate transformation that takes an arbitrary metric of form (1.1.7) over to the simpler form (1.1.8), (see (3.1.4)-(3.1.8) in Chapter 3 below, and [35, 12]). In these notes we deal exclusively with metrics in the form (1.1.8).

The geodesics of a metric are paths $x(s)$ of extremal length, determined by the geodesic equation

$$\frac{d^2 x^i}{ds^2} = \Gamma^i_{jk} \frac{dx^j}{ds} \frac{dx^k}{ds}, \quad (1.1.9)$$

where the so called Christoffel symbols or connection coefficients Γ^i_{jk} are defined by

$$\Gamma^i_{jk} = \frac{1}{2} g^{\sigma i} \{-g_{jk,\sigma} + g_{\sigma j,k} + g_{k\sigma,j}\}. \quad (1.1.10)$$

(Here ",k" denotes the classical derivative in direction x^k.) The Christoffel symbols Γ^i_{jk} are the central objects of differential geometry that do not transform like a tensor. Indeed, they fail to be tensorial by exactly the amount required to convert coordinate differentiation of vector components into a tensorial operation. That is, for a vector field Y, let Y^i denote the x^i-component of Y. The covariant derivative ∇ is defined by

$$\nabla_{\frac{\partial}{\partial x^\sigma}} Y = Z,$$

where Z defines a vector field with x-components

$$Z^i = Y^i_{;\sigma} \equiv \frac{\partial Y^i}{\partial x^\sigma} - \Gamma^i_{jk} X^j X^k. \quad (1.1.11)$$

For arbitrary vector fields X and Y, one defines the covariant derivative $\nabla_X Y$ by

$$\nabla_X Y = X^\sigma \nabla_{\frac{\partial}{\partial x^\sigma}} Y \equiv X^\sigma Y^i_{;\sigma} \frac{\partial}{\partial x^i}.$$

We say that a vector field Y is parallel along a curve whose tangent vector is X if

$$\nabla_X Y = 0,$$

all along the curve. It follows that the covariant derivative $\nabla_X Y$ measures the rate at which the vector field Y diverges from the parallel translation of Y in the direction of X. In a similar fashion, one can define the covariant derivative ∇T of any (k, l) tensor T as the $(k, l+1)$ tensor with components

$$T^{i_1, \ldots, i_k}_{j_1, \ldots, j_l; \sigma}.$$

For example, for a $(1, 1)$ tensor T,

$$T^i_{j;\sigma} = T^i_{j,\sigma} - \Gamma^i_{\tau\sigma} T^\tau_j + \Gamma^\tau_{j\sigma} T^i_\tau. \tag{1.1.12}$$

More generally, to compute ∇T for a (k, l) tensor T, include a negative term for every contravariant index, (contract the index with Γ as above), and a positive term (as above) for every covariant index in T. We say that T is parallel along a curve with tangent vector X if $\nabla_X T = 0$ all along the curve. It follows that $\nabla_X T$ measures the rate at which T diverges from the parallel translation of T in direction X. For a $(2, 0)$ tensor T we define the covariant divergence of T to be the vector field defined by

$$divT = T^{i\sigma}_{;\sigma} \frac{\partial}{\partial x^i}. \tag{1.1.13}$$

The covariant derivative commutes with contraction and the raising and lowering of indices, [35], and by (1.1.12), ∇ reduces to the classical derivative at any point where the Christoffel symbols Γ^i_{jk} vanish.

It follows from (1.1.10) that $\Gamma^i_{jk} = 0$ at a point in a coordinate system where $g_{ij,k} = 0$, all $i, j, k = 0, \ldots, 3$. The existence of such coordinate frames at a point follows directly from the fact that the metric components g_{ij} are smoothly varying, and transform like a symmetric bilinear form under coordinate transformation. If in addition, $g_{ij} = diag(-1, 1, 1, 1)$, then such a coordinate system is said to be locally inertial, or locally Lorentzian at the point. The notion of geodesics and parallel translation have a very natural physical interpretation in General Relativity in terms of the locally inertial coordinate frames. Indeed, General Relativity makes contact with (the flat spacetime theory of) Special Relativity by identifying the locally Lorentzian frames at a point as the "locally non-rotating" inertial coordinate systems in which spacetime behaves as if it were locally flat. Thus physically, the non-rotating vector fields carried by an observer in freefall should be the vector fields that are *locally constant in the locally inertial coordinate frames defined at each point along the curve.* But since $\Gamma^i_{jk} = 0$ at the center of a locally inertial coordinate system, it follows from (1.1.11) that a vector field is parallel translated along a curve, (in the sense that $\nabla_X Y = 0$ along a curve), if and only if its components are (locally) constant in the locally inertial coordinate frames defined at each point along the curve. Thus we see that the non-rotating vector fields carried by an observer in freefall are exactly the vectors that are parallel transported by the unique symmetric connection (1.1.10) determined by the gravitational metric g. Similarly, the geodesics of the metric g are just the curves that are "locally straight lines" in the locally inertial coordinate frames.

The fundamental tenet of General Relativity is the principle that there is no apriori global inertial coordinate system on spacetime. Rather, in General Relativity, inertial coordinate systems are *local* properties of spacetime in the sense that they change from point to point. For example, if there were a global Newtonian *absolute space*, then there would exist global coordinate systems in which freefalling objects do not accelerate, and any two such coordinate systems would be related by transformations from the 10 parameter Galilean Group–the set of coordinate transformations that do not introduce accelerations. In Special Relativity, the existence of absolute space would presume the existence of global coordinate systems related by the transformations of Special Relativity; that is, in Special Relativity, the 10 parameter Poincare group replaces the 10 parameter Galilean Group as the set of transformations that introduce no accelerations. The Poincare Group is obtained from the Galilean group by essentially replacing Euclidean translation in time by Lorentz transformations, and this accounts for time dilation. The spacetime metric can then be viewed as a book-keeping device for keeping track of the location of the local inertial reference frames as they vary from point to point in a given coordinate system–the metric locates the local inertial frames at a given point as those coordinate systems that diagonalize the metric at that point, $g_{ij} = diag(-1, 1, 1, 1)$, such that the derivatives of the metric components also vanish at the point. Thus, the earth moves "unaccelerated" in each local inertial frame, but these frames change from point to point, thus producing apparent accelerations in a global coordinate system in which the metric is not everywhere diagonal. The fact that the earth moves in a periodic orbit around the Sun is proof that there is no coordinate system that globally diagonalizes the metric, and this is an expression of the fact that gravitational fields produce nonzero spacetime curvature. Indeed, in an inertial coordinate frame, when a gravitational field is present, one cannot in general eliminate the second derivatives of the metric components at a point by any coordinate transformation, and the nonzero second derivatives of the metric that cannot be eliminated, represent the gravitational field. These second derivatives are measured by the Riemann Curvature Tensor associated with the Riemannian metric g.

Riemann introduced the curvature tensor in his inaugural lecture of 1854. In this lecture he solved the longstanding open problem of describing curvature in surfaces of dimension higher than two. Although the curvature tensor was first developed for positive definite "spatial" metrics, Einstein accounted for time dilation by letting Lorentz transformations play the role of rotations in Riemann's theory, and except for this, Riemann's theory carries over essentially unchanged. The Riemann Curvature Tensor $R^i_{jkl}(\mathbf{x})$ is a quantity that involves second derivatives of $g_{ij}(\mathbf{x})$, but which transforms like a tensor under coordinate transformation; that is, the components transform like a sort of four component version of a vector field, even though vector fields are constructed essentially from first order derivatives. The connection between General Relativity and geometry can be summarized in the statement that the Riemann Curvature Tensor associated with the metric g gives an invariant description of gravitational accelerations. The components of the Riemann Curvature Tensor are given in terms of the Christoffel symbols by the formula, [34],

$$R^i_{jkl} = \Gamma^i_{jl,k} - \Gamma^i_{jk,l} + \left\{ \Gamma^\sigma_{jl}\Gamma^i_{\sigma k} - \Gamma^\sigma_{jk}\Gamma^i_{\sigma l} \right\}. \tag{1.1.14}$$

One can interpret this as a "curl" plus a "commutator".

1.2 Introduction to the Einstein Equations

Once one makes the leap to the idea that the inertial coordinate frames change from point to point in spacetime, one is immediately stuck with the idea that, since our non-rotating inertial frames here on earth are also non-rotating with respect to the fixed stars, the stars must have had something to do with the determination of our non-accelerating reference frames here on earth, (Mach's Principle). Indeed, not every Lorentzian metric can describe a gravitational field, which means that gravitational metrics must satisfy a constraint that describes how inertial frames at different points of spacetime interact and evolve. In Einstein's theory of gravity, this constraint is given by the Einstein gravitational field equations. These field equations were first introduced by Albert Einstein in 1915 after nine years of struggle.

The Einstein equations can be written in the compact form

$$G = \kappa T. \tag{1.2.1}$$

Here G denotes the Einstein curvature tensor, T the stress energy tensor, (the source of the gravitational field), and κ is a universal constant. In a given coordinate system x, the field equations (1.2.1) take the component form

$$G_{ij}(x) = \kappa T_{ij}(x), \tag{1.2.2}$$

where

$$G_{ij} \equiv R^{\sigma}_{\ i\sigma j} - \frac{1}{2} R^{\sigma\tau}_{\ \sigma\tau} g_{ij}, \qquad (1.2.3)$$

denote the x-components of the Einstein curvature tensor, and T_{ij} the x-components of the stress energy tensor. We let $0 \leq i, j \leq 3$ refer to components in a given coordinate system, and again we assume the Einstein summation convention whereby repeated up-down indices are assumed to be summed from 0 to 3. The components of the stress energy tensor give the energy density and i-momentum densities and their fluxes at each point of spacetime. When the sources are modeled by a perfect fluid, T is given (in contravariant form) by

$$T^{ij} = (\rho + p)w^i w^j + p g^{ij}, \qquad (1.2.4)$$

where $\mathbf{w}$ denotes the unit 4-velocity vector of the fluid, (the tangent vector to the world line of the fluid particle), ρ denotes the energy density, (as measured in the inertial frame moving with the fluid), and p denotes the fluid pressure. The four velocity $\mathbf{w}$ has components $w^i = \frac{dx^i}{ds}$ when the fluid particle traverses a (timelike) path $x(s)$ in x-coordinates, and s is taken to be the arclength parameter (1.1.1) determined by the gravitational metric g. It follows that $\mathbf{w}$ is a unit timelike vector relative to g, and thus only three of the four components of $\mathbf{w}$ are independent.

The constant κ in (1.2.1) is determined by the principle that the theory should incorporate Newton's theory of gravity in the limit of low velocities and weak gravitational fields, (correspondence principle). This leads to the value

$$\kappa = 8\pi\mathcal{G}/c^4.$$

Here c denotes the speed of light and $\mathcal{G}$ denotes Newton's gravitational constant. Newton's constant first appears in the inverse square force law

$$Force = M\mathbf{a} = -\mathcal{G}\frac{MM_0}{r^3}\mathbf{r}. \qquad (1.2.5)$$

In (1.2.5), M is the mass of a planet, M_0 is the mass of the sun, and $\mathbf{r}$ is the position vector of the planet relative to the center of mass of the system. The Newton law (1.2.5) starts looking like it isn't really a "fundamental law" once one verifies that the inertial mass M on the LHS of (1.2.5) is equal to the gravitational mass M on the RHS of (1.2.5), (Equivalence Principle). In this case, M cancels out, and then (1.2.5) (remarkably) becomes more like a law about accelerations than a law about "forces". That is, once M cancels out, the force law (1.2.5) is independent of any properties of the object (planet) whose motion it purports to describe. Thus, in Newton's theory, the "gravitational force", which is different on different objects of different masses, miraculously adjusts itself perfectly so that every object, (subject to the same initial conditions), traverses exactly the same path. Thus Einstein was led to suspect that the Newtonian gravitational force was some sort of artificial device, and that the fundamental objects of the gravitational field were the "freefall paths", not the forces. From this point of view, the field equations (1.2.1) are more natural than (1.2.5) because they are, at the start, equations for the gravitational metric, and the gravitational metric fundamentally describes the paths of "freefalling" objects by means of the geodesic equation of motion (1.1.9)—which just expresses "local non-acceleration in locally inertial coordinate frames". In Newton's theory of gravity, the non-rotating frames here on earth are aligned with the stars because there is a global inertial coordinate system that connects us. In contrast, according to the modern theory of cosmology, which is based on Einstein's theory of gravity, the non-rotating inertial frames here on earth are aligned with the stars because the FRW metric (1.2.1) maintains this alignment, and (1.2.1) solves the Einstein equations for an appropriate choice of $R(t)$. (This is still a bit unsatisfying!)

In the limit that a finite set of point masses tends to a continuous mass distribution with density ρ, Newton's force law is replaced by the Poisson equation for the gravitational potential ϕ,

$$-\Delta\phi = 4\pi\mathcal{G}\rho. \tag{1.2.6}$$

Indeed, in the case of a compactly supported density $\rho(x)$, one can use the fundamental solution of Laplacian to write the solution of (1.2.6) as

$$\phi(x) = \int_{R^3} \frac{\mathcal{G}}{|x - y|}\rho(y)d^3y, \tag{1.2.7}$$

so the Newtonian acceleration at a point x is given by

$$\mathbf{a} = -\nabla\phi = \int_{R^3} \frac{\mathcal{G}}{|x - y|^3}(x - y)\rho(y)d^3y. \tag{1.2.8}$$

Thus we recover (1.2.5) from (1.2.8) by approximating ρ in (1.2.8) by a finite number of point masses.

The Einstein equations play the same role in General Relativity that the Poisson equation (1.2.6) plays in the Newtonian theory of gravity—except there is a very significant difference: the Poisson equation determines the (scalar) gravitational potential ϕ given the mass density ρ, but in Newton's theory this must be augmented by some system of conservation laws in order to describe the time evolution of the mass density ρ as well. For example, if we assume that the density evolves according to a perfect fluid with pressure p and 3-velocity $\mathbf{v}$, then the coupling of Newton's law of gravity with the Euler equations for a perfect fluid leads to the Euler-Poisson system

$$\rho_t + div(\rho\mathbf{v}) = 0,$$
$$(\rho v^i)_t + div(\rho v^i\mathbf{v} + p\mathbf{e}^i) = -\rho\nabla\phi, \tag{1.2.9}$$
$$-\Delta\phi = 4\pi\mathcal{G}\rho.$$

The first four equations are the compressible Euler equations with the gravitational forcing term on the RHS. The first equation, the continuity equation, expresses conservation of mass, the next three express conservation of i-momentum, $i = 1, 2, 3$, (for a perfect fluid this really says that the time rate of change of momentum is equal to the sum of the force of the pressure gradient plus the force of the gravitational field; e^i denotes the i'th unit vector in R^3), and the last equation expresses the continuum version of Newton's inverse square force law. Note that for the fluid part of (1.2.9), information propagates at the sound (and shock) speeds, but the gravitational potential ϕ is updated "instantaneously", depending only on the density $\rho(x, t)$, according to the formula (1.2.7). In contrast, for Einstein's theory of gravity, the time evolution of the gravitational metric is determined simultaneously with the time evolution of the sources through system (1.2.1), and all of the components of the stress tensor directly influence the components of the gravitational field g_{ij}. This principle is the basis for the discovery of the Einstein equations. Indeed, since the 0-column of the stress-energy tensor (1.2.4) gives the energy and momentum densities, and the i-column gives the corresponding i-fluxes, (in the relativistic sense), it follows that conservation of energy-momentum in curved spacetime reduces to the statement

$$Div(T) = 0, \qquad (1.2.10)$$

where (capital) Div denotes the covariant divergence for the metric g, so that it agrees with the ordinary divergence in each local inertial coordinate frame. In this way equations (1.2.10) reduce to the relativistic compressible Euler equations in flat Minkowski spacetime. Since the covariant derivative depends on the metric components, the conservation equation (1.2.10) is essentially coupled to the equation for the gravitational field g. But the stress tensor T is symmetric, $T_{ij} = T_{ji}$, and so the tensor on the LHS of (1.2.2) must also be symmetric. Therefore the Einstein equations (1.2.2) supply ten independent equations in the ten independent unknown metric components g_{ij}, together with the four independent functions among ρ and the unit vector field $\mathbf{w}$. (For example, assume p is determined by ρ through an equation of state of the form $p = p(\rho)$.) But (1.2.2) assumes no coordinate system, and thus in principle we are free to give four further relations that tie the components of G and T to the coordinate system. This leaves ten equations in ten unknowns, and thus there are no further constraints allowable to couple system (1.2.1) to the conservation laws (1.2.10). The only way out is to let (1.2.10) follow as an identity from (1.2.1), and this determines the LHS of (1.2.1), namely, the Einstein tensor G_{ij} is the simplest tensor constructable from R^i_{jkl} such that (1.2.10) follows identically from the Bianchi identities of Riemannian geometry, ($R^i_{j[kl,m]} = 0$, where $[kl, m]$ denotes cyclic sum, c.f., [35]). [4] Thus,

---

[4] This is the simplest known route to the field equations (1.2.1). Of course, since (1.2.1) represents a new starting point, it follows that there must be a "conceptual leap" at some stage of any "derivation" of (1.2.1).

the simplest and most natural field equations of form (1.2.1) are uniquely determined by the equation count, [35]. The next simplest tensor for the LHS of (1.2.1) that meets (1.2.10) is

$$G_{ij} + \Lambda g_{ij},$$

where Λ is the famous *cosmological constant* first identified by Albert Einstein in his attempt to construct a cosmological model that was static, c.f. [35]. In these notes we always assume $\Lambda = 0$. One can show that in the limit of low velocities and weak gravitational fields, the equations (1.2.10) reduce to the first four equations of (1.2.9), and the $(0,0)$ component of the Einstein equations (1.2.2) reduces to the Poisson equation (1.2.6), thus fixing the choice $\kappa = 8\pi\mathcal{G}/c^4$, [35]. This establishes the correspondence of Einstein's theory of gravity with the Newtonian theory.

To summarize, in Einstein's theory of gravity, based on (1.2.1), the conservation of energy and momentum (1.2.10) are not imposed, but follow as differential identities from the field equations (1.2.1). In a specified system of coordinates, (1.2.1) determines a hyperbolic system of equations that simultaneously describes the time evolution and interaction of local inertial coordinate frames, as well as the time evolution of the fluid according to (1.2.10). Since GR is coordinate independent, we can always view the time evolution (1.2.1) in local inertial coordinates at any point in spacetime, in which case (1.2.10) reduces to the classical relativistic Euler equations at the point. This tells us that, heuristically, shock waves must form in the time evolution of (1.2.1) because one could in principle drive a solution into a shock while in a neighborhood where the equations remained a small perturbation of the classical Euler equations. (This is much easier to say than to demonstrate rigorously, and as far as we know, such a demonstration has not been given.)

In these notes, we assume that shock waves are as fundamental to the time evolution of solutions of the Einstein equations for a perfect fluid, as they are for the time evolution of the classical compressible Euler equations (1.2.9). At a shock wave, the fluid variables ρ, $\mathbf{w}$ and p are discontinuous. Notice that (1.2.1) implies that the Einstein curvature tensor G will be discontinuous at any point where T is discontinuous. Since G involves second derivatives of the metric tensor g, the only way (1.2.1) can hold in the classical pointwise a.e. sense at the shock is if the component functions g_{ij} are continuously differentiable at the shock, with bounded derivatives on either side; that is, if $g_{ij} \in C^{1,1}$. Thus we expect from (1.2.1) that the spacetime metric g should be $C^{1,1}$ at shock waves. However, we now show that for a spherically symmetric metric in standard Schwarzschild coordinates (1.1.8), the best one can expect is that $g \in C^{0,1}$.

1.3 The Simplest Setting for Shock Waves

In these notes we assume the simplest possible setting for shock wave propagation in General Relativity: namely, the case of a spherically symmetric metric in standard Schwarzschild coordinates (1.1.8), assuming a perfect fluid (1.2.4) with equation of state

$$p = \sigma^2 \rho, \quad 0 < \sigma < c, \tag{1.3.1}$$

where σ, the sound speed, is assumed to be constant.[5] Using MAPLE to put the metric ansatz (1.1.8) into the Einstein equations (1.2.1) produces the following system of four coupled partial differential equations, (c.f. (3.20)-(3.23) of [12]),

$$\frac{A}{r^2 B} \left\{ r \frac{B'}{B} + B - 1 \right\} = \kappa A^2 T^{00}, \tag{1.3.2}$$

$$-\frac{B_t}{rB} = \kappa ABT^{01}, \tag{1.3.3}$$

$$\frac{1}{r^2} \left\{ r \frac{A'}{A} - (B - 1) \right\} = \kappa B^2 T^{11}, \tag{1.3.4}$$

$$-\frac{1}{rAB^2} \{ B_{tt} - A'' + \Phi \} = \frac{2\kappa r}{B} T^{22}, \tag{1.3.5}$$

where the quantity Φ in the last equation is given by,

$$\Phi = -\frac{BA_t B_t}{2AB} - \frac{B}{2} \left(\frac{B_t}{B} \right)^2 - \frac{A'}{r} + \frac{AB'}{rB}$$
$$+ \frac{A}{2} \left(\frac{A'}{A} \right)^2 + \frac{A}{2} \frac{A'}{A} \frac{B'}{B}.$$

Here "prime" denotes $\partial/\partial r$, "dot" denotes $\partial/\partial t$, $\kappa = \frac{8\pi \mathcal{G}}{c^4}$ is again the coupling consant, $\mathcal{G}$ is Newton's gravitational constant, c is the speed of light, T^{ij}, $i, j = 0, ..., 3$ are the components of the stress energy tensor, and $A \equiv A(r, t)$, $B \equiv B(r, t)$ denote the components of the gravitational metric tensor (1.1.8) in standard Schwarzschild coordinates $\mathbf{x} = (x^0, x^1, x^2, x^3) \equiv (t, r, \theta, \phi)$. The mass function M is defined through the identity

---

[5] This simplifying assumption, as well as insuring that wave speeds are bounded by the speed of light for arbitrarily strong shock waves, also prevents the formation of vacuum states, and allows us to exploit special properties of the relativistic compressible Euler equations derived in Chapter 2. The results of Chapter 3 regarding the weak equivalence of the Einstein equations with a system of conservation laws with time dependent sources, hold for general p.

$$B = \left(1 - \frac{2M}{r}\right)^{-1}, \tag{1.3.6}$$

and $M \equiv M(r,t)$ is interpreted as the mass inside radius r at time t. In terms of the variable M, it follows directly that equations (1.3.2) and (1.3.3) are equivalent to

$$M' = \tfrac{1}{2}\kappa r^2 A T^{00}, \tag{1.3.7}$$

and

$$\dot{M} = -\tfrac{1}{2}\kappa r^2 A T^{01}, \tag{1.3.8}$$

respectively. Using the perfect fluid assumption (1.2.4), the components T^{ij} satisfy

$$T^{00} = \frac{1}{A} T_M^{00}, \tag{1.3.9}$$

$$T^{01} = \frac{1}{\sqrt{AB}} T_M^{01}, \tag{1.3.10}$$

$$T^{11} = \frac{1}{B} T_M^{11}, \tag{1.3.11}$$

where T_M^{ij} denote the components of T in flat Minkowski spacetime. When $p = \sigma^2 \rho$, the components of T_M are given by

$$T_M^{00} = \frac{c^4 + \sigma^2 v^2}{c^2 - v^2} \rho, \tag{1.3.12}$$

$$T_M^{01} = \frac{c^2 + \sigma^2}{c^2 - v^2} cv\rho, \tag{1.3.13}$$

$$T_M^{11} = \frac{v^2 + \sigma^2}{c^2 - v^2} \rho c^2, \tag{1.3.14}$$

c.f., [27, 12]. Here v, taken in place of $\mathbf{w}$, denotes the fluid velocity as measured by an observer fixed with respect to the radial coordinate r. It follows from (1.3.7) together with (1.3.12)-(1.3.14) that, if $r \geq r_0 > 0$, then

$$M(r,t) = M(r_0,t) + \frac{\kappa}{2} \int_{r_0}^{r} T_M^{00}(r,t) r^2 \, dr; \tag{1.3.15}$$

it follows from (1.2.2) together with (1.3.12)-(1.3.14), that the scalar curvature R is proportional to the density,

$$R = (c^2 - 3\sigma^2)\rho; \tag{1.3.16}$$

and it follows directly from (1.3.12)-(1.3.14) that

$$|T_M^{01}| < T_M^{00}, \tag{1.3.17}$$

$$\frac{\sigma^2}{c^2+\sigma^2}T_M^{00} < T_M^{11} < T_M^{00}. \tag{1.3.18}$$

Equations (1.3.1)-(1.3.18) define the simplest possible setting for shock wave propagation in the Einstein equations.

1.4 A Covariant Glimm Scheme

When shock waves are present in solutions of (1.3.2)-(1.3.5), T is discontinuous, and so it follows from (1.3.2) and (1.3.4) that A and B will be at best Lipschitz continuous, and so equation (1.3.5) can only hold in the weak sense. We prove the existence of such weak solutions of the Einstein equations in Chapter 4. In Chapter 3 we show that when A and B are Lipschitz continuous functions of (t, r), and T is bounded in L^∞, system (1.3.2)-(1.3.5) is weakly equivalent to the system obtained by replacing (1.3.3) and (1.3.5) by the system $Div T = 0$ in the form,

$$\left\{T_M^{00}\right\}_{,0} + \left\{\sqrt{\frac{A}{B}}T_M^{01}\right\}_{,1} = -\frac{2}{x}\sqrt{\frac{A}{B}}T_M^{01}, \tag{1.4.1}$$

$$\left\{T_M^{01}\right\}_{,0} + \left\{\sqrt{\frac{A}{B}}T_M^{11}\right\}_{,1} = -\frac{1}{2}\sqrt{\frac{A}{B}}\left\{\frac{4}{x}T_M^{11} + \frac{(B-1)}{x}(T_M^{00} - T_M^{11})\right. \tag{1.4.2}$$

$$\left. +2\kappa x B(T_M^{00}T_M^{11} - (T_M^{01})^2) - 4xT^{22}\right\}.$$

(We use x in place of r when the equations are expressed as a system of conservation laws.) This is a nice formulation of $Div T = 0$ because the conserved variables $u = (T_M^{00}, T_M^{01})$ are the Minkowski energy and momentum densities, (c.f. (1.3.12), (1.3.13)), and thus do not depend on the metric components $\mathbf{A} \equiv (A, B)$. Note that all terms involving time derivatives of A and B have cancelled out from the RHS of (1.4.1), (1.4.2), a nice feature of this choice of variables. (Note also that there is no $\dot{A}$ equation among (1.3.2)-(1.3.5) with which to eliminate $\dot{A}$ terms, so some change of variables is required to eliminate such terms from $Div T = 0$, c.f. (3.3.3), (3.3.4) below.) What results is a system of conservation laws with source terms which can be written in the compact form

$$u_t + f(\mathbf{A}, u)_x = g(\mathbf{A}, u, x), \tag{1.4.3}$$

$$\mathbf{A}' = h(\mathbf{A}, u, x), \tag{1.4.4}$$

where the first equation is (1.4.1),(1.4.2), and the second equation is (1.3.2),(1.3.4), so that

$$u = (T_M^{00}, T_M^{01}) \equiv (u^0, u^1),$$
$$\mathbf{A} = (A, B),$$
$$f(\mathbf{A}, u) \equiv (f^0, f^1) = \sqrt{\frac{A}{B}} \, (T_M^{01}, T_M^{11}),$$

and $g = (g^0, g^1)$ is determined from the RHS of (1.4.1), (1.4.2), while $h = (h^0, h^1)$ is determined from the RHS of (1.3.4), (1.3.2) upon solving for (A', B'), respectively, (c.f. (4.1.12)-(4.1.13) below).

The existence theory in Chapter 4 is based on a fractional step Glimm scheme first introduced in [13]. The fractional step method employs a Riemann problem step, (the Riemann problem is discussed in Chapter 2, c.f. [26]), that simulates the source free conservation law $u_t + f(\mathbf{A}, u)_x = 0$, $(\mathbf{A} \equiv Const)$, followed by an ODE step that accounts for the sources present in both f and g. The idea for the numerical scheme is to stagger discontinuities in the metric with discontinuities in the fluid variables so that the Riemann problem step, as well as the ODE step of the method, are both generated in grid rectangles on which the metric components $\mathbf{A} = (A, B)$, (as well as x), are constant. The equation $A' = h(\mathbf{A}, u, x)$ is solved at the end of each time step, and Glimm's method of random choice is employed to re-discretize at the start of the next time step. (See Section 4.3 for a precise definition of the fractional step scheme.) Part of the proof of existence involves showing that the ODE step $u_t = g(\mathbf{A}, u, x) - \nabla_{\mathbf{A}} f \cdot \mathbf{A}'$, with h substituted for $\mathbf{A}'$, accounts for both the source term g, as well as the *effective sources* that are due to the discontinuities in the metric components at the boundaries of the grid rectangles.

Because only the flux f in the Riemann problem step depends on $\mathbf{A}$, it follows that the only effect of the metric on the Riemann problem step of the method is to change the wave speeds, but except for this, the solution of the Riemann problem in each grid rectangle agrees with the solution in flat Minkowski spacetime. At this stage, we can apply the estimates which we obtain in Chapter 2 for the problem $divT = 0$ in flat Minkowski spacetime–the total variation of $\ln \rho_{\Delta x}(\cdot, t)$ is non-increasing on the Riemann problem step of the method. Thus to obtain compactness of approximate solutions, one needs only show that the increase in the total variation of $\ln \rho_{\Delta x}(\cdot, t)$ produced by the ODE step and the imposition of the constraints, is order Δx. The main technical problem in achieving this is to keep track of the order of choice of constants and to show that the total mass $M_\infty = \frac{\kappa}{2} \int_{r_0}^{\infty} \rho r^2 \, dr$ is bounded. The problem is that, in the estimates, the growth of ρ depends on M and the growth of M depends on ρ, and M is defined by a *non-local* integral. Thus, an error estimate of order Δx for $\Delta \rho$ after one time step, is not sufficient to bound the total mass M_∞ after one time step.

We conclude the Introduction by returning to our theme that the fractional step method of Chapter 4 can be viewed as a *locally inertial* version of Glimm's method in the sense that it exploits the locally flat character of spacetime. To see this, note that the Riemann problem step solves the equations $u_t + f(\mathbf{A}, u)_x = 0$ inside grid rectangles $\mathcal{R}_{ij}$ where $\mathbf{A} \equiv Const$. Thus each grid rectangle is a local inertial reference frame because the metric is flat when $\mathbf{A} \equiv Const$. The boundaries between these local inertial reference frames are the discontinuities that appear along the top, bottom and both sides of the grid rectangles. These discontinuities are determined through the ODE step $u_t = g(\mathbf{A}, u, x) - \nabla_{\mathbf{A}} f \cdot \mathbf{A}'$, followed by the random choice and the imposition of the constraint $A' = h(\mathbf{A}, u, x)$. The term $-\nabla_{\mathbf{A}} f \cdot \mathbf{A}'$ on the RHS of the ODE step accounts for the discontinuities in $\mathbf{A}$ along the sides of the grid rectangles $\mathcal{R}_{ij}$, and the term g in the ODE step, together with the constraint $A' = h(\mathbf{A}, u, x)$ at the end of each timestep, both account for the discontinuities in $\mathbf{A}$ at the top and bottom of each $\mathcal{R}_{ij}$. It follows that once the convergence of an approximate solution is established, one can just as well replace the true approximate solution by the solution of the Riemann problem in each grid rectangle $\mathcal{R}_{ij}$–the two differ by only order Δx. The resulting appoximation scheme converges to a weak solution of the Einstein equations, and has the property that it solves the compressible Euler equations exactly in local inertial coordinate frames, (grid rectangles), and accounts for the transformations between neighboring coordinate frames by discontinuities at the coordinate boundaries.

We conclude by reiterating that the fractional step Glimm method introduced in Chapter 4 for the analysis of the initial value problem, (material taken from [13]), is natural for the Einstein equations because it is a locally inertial method that exploits the locally flat character of spacetime. Such a method *requires* that shock waves be accounted for because the Riemann problem is essentially incomplete without shock waves.

We begin Chapter 2 with a discussion of the relativistic compressible Euler equations $div T = 0$ in flat Minkowski spacetime.

2

The Initial Value Problem in Special Relativity

2.1 Shock Waves in Minkowski Spacetime

We consider the relativistic equations for a perfect fluid

$$divT = 0, \tag{2.1.1}$$

in flat Minkowski spacetime,

$$ds^2 = \eta_{ij}dx^i dx^j = -d(ct)^2 + d(x^1)^2 + d(x^2)^2 + d(x^3)^2, \tag{2.1.2}$$

where

$$T^{ij} = (p + \rho c^2)w^i w^j + p\eta^{ij}, \tag{2.1.3}$$

denotes the stress-energy tensor for the fluid. Recall that in Minkowski spacetime,

$$divT \equiv T^i_{j,i} \tag{2.1.4}$$

where again we assume summation over repeated up-down indices, ", i" denotes differentiation with respect to the variable x^i, and in general all indices run from 0 to 3 with $x^0 \equiv ct$. In (2.1.3), c denotes the speed of light, (we take $c = 1$ when convenient), p the pressure, $\mathbf{w} = (w^0, ..., w^3)$ the 4−velocity of the fluid particle, ρ the mass-energy density, and $\eta^{ij} \equiv \eta_{ij} \equiv diag(-1, 1, 1, 1)$.

In this section we study the initial value problem for (2.1.1) in a two dimensional spacetime (x^0, x^1), so that ρ and $\mathbf{w}$ are unknown functions of (x^0, x^1), and

$$\eta^{ij} = \begin{bmatrix} -1 & 0 \\ 0 & 1 \end{bmatrix}. \tag{2.1.5}$$

Under these assumptions the stress-energy tensor (2.1.3) takes the form

$$T^{ij} = \begin{bmatrix} (p + \rho c^2)w^0 w^0 - p & (p + \rho c^2)w^0 w^1 \\ (p + \rho c^2)w^0 w^1 & (p + \rho c^2)w^1 w^1 + p \end{bmatrix}. \tag{2.1.6}$$

For our theorem we assume that p and ρ satisfy an equation of state of the form

$$p = \sigma^2 \rho, \tag{2.1.7}$$

where σ^2, the sound speed, is taken to be constant, $\sigma < c$. In particular, when $\sigma^2 = c^2/3$, (2.1.7) gives the important relativistic case $p = (c^2/3)\rho$ discussed above. Since the background metric is the flat Minkowski metric η_{ij}, the increment of proper time τ, (Minkowski arclength), along a curve is given by the formula

$$d(\xi)^2 = -\eta_{ij}dx^i dx^j, \tag{2.1.8}$$

where we use the notation $\xi = c\tau$. In this way the coordinate time t and the proper time τ have the dimensions of time, while x^0 and ξ have the dimensions of length. Since $w^i = dx^i/d(c\tau)$, (where differentiation is taken along a particle path), defines the dimensionless velocity of the fluid, we must have

$$w^0 = \sqrt{1 + (w^1)^2}. \tag{2.1.9}$$

Thus letting $w \equiv w^1$, the equations we consider are

$$\frac{\partial}{\partial x^0}\{(p + \rho c^2)(1 + w^2) - p\} + \frac{\partial}{\partial x^1}\{(p + \rho c^2)w\sqrt{1 + w^2}\} = 0, \tag{2.1.10}$$

$$\frac{\partial}{\partial x^0}\{(p + \rho c^2)w\sqrt{1 + u^2}\} + \frac{\partial}{\partial x^1}\{(p + \rho c^2)w^2 + p\} = 0,$$

together with the initial data

$$\rho(0, x^1) = \rho_0(x^1), \quad w(0, x^1) = w_0(x^1). \tag{2.1.11}$$

The equations (2.1.10) form a system of nonlinear hyperbolic conservation laws in the sense of Lax [17]. Thus if one seeks global (in time) solutions, then due to the formation of shock waves, one must extend the notion of solution, in the usual way, [26], in order to admit as solutions such discontinuous functions.

In the classical limit, the relativistic system (2.1.10) reduces to the classical version of the compressible Euler equations. In order to observe this correspondence throughout, we set $x \equiv x^1$, choose $\mathbf{x} \equiv (x, t)$ as the independent variables, and replace w in system (2.1.10) in favor of its expression in terms of the classical coordinate velocity $v \equiv dx/dt$ of the particle paths of the fluid. To accomplish this, note that by (2.1.9),

$$\frac{dt}{d\tau} = \frac{dx^0}{d\xi} = \sqrt{1 + w^2},$$

so we can write

$$v = \frac{dx^1}{dt} = \frac{dx^1}{d\xi}\frac{d\xi}{dt} = cw\frac{1}{\sqrt{1 + w^2}}, \tag{2.1.12}$$

which solving for w gives

$$w = v/\sqrt{c^2 - v^2}. \qquad (2.1.13)$$

The mapping $w \rightarrow v$ in (2.1.13) defines a smooth $1 - 1$ mapping from $(-\infty, +\infty)$ to $(-c, c)$, and so there is no loss of generality in taking v as the state variable instead of w.

Now writing system (2.1.10) in terms of ρ and v and multiplying the first equation by $1/c$, we obtain the general system

$$\frac{\partial}{\partial t}\left\{(\frac{p + \rho c^2}{c^2})\frac{v^2}{c^2 - v^2} + \rho\right\} + \frac{\partial}{\partial x}\left\{(p + \rho c^2)\frac{v}{c^2 - v^2}\right\} = 0,$$

$$(2.1.14)$$

$$\frac{\partial}{\partial t}\left\{(p + \rho c^2)\frac{v}{c^2 - v^2}\right\} + \frac{\partial}{\partial x}\left\{(p + \rho c^2)\frac{v^2}{c^2 - v^2} + p\right\} = 0.$$

Restricting to the case $p = \sigma^2 \rho$, (2.1.14) reduces to

$$\frac{\partial}{\partial t}\left\{\rho[\left(\frac{\sigma^2 + c^2}{c^2}\right)\frac{v^2}{c^2 - v^2} + 1]\right\} + \frac{\partial}{\partial x}\left\{\rho[(\sigma^2 + c^2)\frac{v}{c^2 - v^2}]\right\} = 0,$$

$$(2.1.15)$$

$$\frac{\partial}{\partial t}\left\{\rho[(\sigma^2 + c^2)\frac{v}{c^2 - v^2}]\right\} + \frac{\partial}{\partial x}\left\{\rho[(\sigma^2 + c^2)\frac{v^2}{c^2 - v^2} + \sigma^2]\right\} = 0,$$

together with the initial conditions

$$\rho(x, 0) = \rho_0(x), \quad v(x, 0) = v_0(x). \qquad (2.1.16)$$

Note that in the limit $c \rightarrow \infty$, the system (2.1.14) reduces to the classical system

$$\rho_t + (\rho v)_x = 0,$$

$$(2.1.17)$$

$$(\rho v)_t + (\rho v^2 + \sigma^2 \rho)_x = 0.$$

The main purpose of this chapter is to prove the following theorem:

Theorem 1 *Let $\rho_0(x)$ and $v_0(x)$ be arbitrary initial data satisfying*

$$Var\{\ln(\rho_0(\cdot))\} < \infty, \qquad (2.1.18)$$

and

$$Var\left\{\ln\left(\frac{c + v_0(\cdot)}{c - v_0(\cdot)}\right)\right\} < \infty, \qquad (2.1.19)$$

where $Var\{f(\cdot)\}$ denotes the total variation of the function $f(x)$, $x \in \mathbf{R}$. Then there exists a bounded weak solution $(\rho(x, t), v(x, t))$ of (2.1.15) satisfying

$$Var\{\ln(\rho(\cdot, t))\} \leq V_0, \qquad (2.1.20)$$

and

$$Var\left\{\ln\left(\frac{c+v(\cdot,t)}{c-v(\cdot,t)}\right)\right\} \leq V_1, \tag{2.1.21}$$

where (2.1.20) and (2.1.21) are Lorentz invariant statements, and V_0 and V_1 are Lorentz invariant constants depending only on the initial total variation bounds assumed in (2.1.18) and (2.1.19). Moreover, the solution is a limit of approximate solutions $(\rho_{\Delta x}, v_{\Delta x})$ which satisfy the "energy inequality"

$$Var\left\{\ln(\rho_{\Delta x}(t+,\cdot))\right\} \leq Var\left\{\ln(\rho_{\Delta x}(s+,\cdot))\right\}, \tag{2.1.22}$$

for all times $0 \leq s \leq t$. The approximate solutions are generated by Glimm's method [10], and converge pointwise a.e., and in L_{loc}^1 at each time, uniformly on bounded subsets of (x,t)-space.

In one space dimension, the total variation of a solution at a fixed time $t \geq 0$ is a natural measure of the total wave strength present in the solution at time t. The existence of a function of the state variables like $\ln(\rho)$ whose total variation in time is non-increasing, is a very special property of system (2.1.15), there being no way to construct such a function for a general 2×2 system of conservation laws [10]. We conjecture that the inequality (2.1.22) is valid for the weak solutions themselves, i.e., that

$$Var\{\ln(\rho(\cdot,t+))\} \leq Var\{\ln(\rho(\cdot,s+))\}, \tag{2.1.23}$$

for all $s \leq t$. Such an inequality would provide a Lorentz invariant monotonicity property of the weak solutions of (2.1.15) that refines the estimate (2.1.20).

To prove Theorem 1, we develop an analysis which parallels that first given by Nishida (1968) in [22] for the classical system (2.1.17). Nishida's result provided the first "big data" global existence theorem for weak solutions of the classical compressible Euler equations, and it remains the only argument for stability of solutions in a derivative norm that applies to arbitrarily large initial data. (Nishida originally treated the Lagrangian formulation of system (2.1.17), [7, 26]. A Lagrangian formulation of the relativistic model can be found in [32].) Theorem 1 shows (surprisingly!) that the ideas of Nishida generalize to the relativistic case (2.1.15) where the equations are significantly more complicated. Indeed, the special properties of the system (2.1.15) that lead to the estimates (2.1.20) and (2.1.21) require *not only* that p be linear in ρ, but are also highly dependent on the specific form of the velocity terms; these appearing in a different and more complicated form in the relativistic equations (2.1.15) than in the classical equations (2.1.17). The technique of Nishida is to analyze solutions via the Glimm difference scheme [10] through an analysis of wave interactions in the plane of Riemann invariants. The main technical point in his analysis involves showing that the shock curves based at different points are congruent in the plane of Riemann invariants. We show that this property carries over to the relativistic case by obtaining a new global parameterization of the shock curves. Of course, in the relativistic case, the shock curves are given by considerably more complicated functions. Our analysis exploits the Lorentz invariance properties of system (2.1.15), and thus we take care to develop the geometric properties of the constructions used in our analysis.

Note that if we non-dimensionalize systems (2.1.15) and (2.1.17) by multiplying through by the appropriate powers of c and replacing $\partial/\partial t$ in terms of $\partial/\partial x^0$, we obtain two systems in the variables ρ and v/c, each parameterized by the dimensionless quantity σ/c. Thus we can say that Theorem 1 and Nishida's result [22] establish a "large data" existence theorem for the two distinct one parameter families of dimensionless systems which correspond to (2.1.15) and (2.1.17). But note also that system (2.1.17) is obtained by taking the limit $c \to \infty$ in (2.1.15), and thus we can obtain the congruence property of the shock curves for (2.1.17) by applying the limit $c \to \infty$ to our formulas for (2.1.15), and in this sense we can view Theorem 1 as a generalization of Nishida's theorem [22]. This is done at the end of Section 2.5.

In his original paper [22], Nishida did not exactly obtain the result that the invariant quantity, $Var\{\ln(\rho)\}$, is non-increasing on approximate solutions. The idea for (2.1.22) in Nishida's case came from Liu, and a similar idea was exploited by Luskin and Temple in [19]; see also [23].

The organization of this Chapter is as follows: In Section 2 we put the problem (2.1.15), (2.1.16) in the context of the general theory of conservation laws, prove the regularity of the mapping from the plane of conserved quantities to the (ρ, v)-plane, and we show that (2.1.20) and (2.1.21) are Lorentz invariant statements. In Section 3 we use the Rankine Hugoniot jump relations to derive the wave speeds λ_i and Riemann invariants for (2.1.14) in the case of a general barotropic equation of state $p = p(\rho)$. In this general setting, we shall also derive necessary and sufficient conditions (on the function $p(\rho)$) for the system (2.1.15) to be strictly hyperbolic and genuinely nonlinear in the sense of Lax [17, 26]. We note that the assumption that wave speeds are bounded by c imposes a linear growth rate on $p(\rho)$ as $\rho \to \infty$, and thus there is a possibilty of losing genuine nonlinearity of the system in this limit. Thus, in Section 3 we describe the properties of what we call the relativistic p-system. In Sections 4-7 we restrict to the case $p = \sigma^2 \rho$ and develop the geometry of the shock curves in the coordinate system of Riemann invariants, solve the Riemann problem, and use the Glimm difference scheme to prove Theorem 1. In the appendix we derive the transformation properties of the Rankine Hugoniot jump relations for general relativistic conservation laws. The analysis applies to arbitrary nonlinear spacetime coordinate transformations in 4-dimensional spacetime with arbitrary Lorentzian spacetime metric. We use this to give a simple derivation of the covariance properties of the characteristics, and the transformation formulas for the characteristic speeds and shock speeds in 2-dimensional Special Relativity.

2.2 The Relativistic Euler Equations as a System of Conservation Laws

In this section we put the problem (2.1.15), (2.1.16) in the context of the general theory of conservation laws, and discuss the Lorentz invariant properties of the system.

The problem (2.1.15) and (2.1.16) is a special case of the initial value problem for a general system of nonlinear hyperbolic conservation laws in the sense of Lax [17, 26],

$$u_t + F(u)_x = 0, \tag{2.2.24}$$

$$u(x,0) = u_0(x), \tag{2.2.25}$$

where in our case

$$u \equiv (u^0, u^1) = \left(\rho[\frac{(\sigma^2 + c^2)}{c^2} \frac{v^2}{c^2 - v^2} + 1], \ \rho(\sigma^2 + c^2)\frac{v}{c^2 - v^2} \right), \tag{2.2.26}$$

and

$$F(u) \equiv (F^0, F^1) = \left(\rho(\sigma^2 + c^2)\frac{v}{c^2 - v^2}, \ \rho[(\sigma^2 + c^2)\frac{v^2}{c^2 - v^2} + \sigma^2] \right). \tag{2.2.27}$$

In order to apply Glimm's method (c.f. Section 7), we need the following result.

Proposition 1 *The mapping $(\rho, v) \to (u^1, u^2) = u$ is $1-1$, and the Jacobian determinant of this mapping is both continuous and nonzero in the region $\rho > 0, |v| < c$.*

Proof: If the mapping were not $1-1$, then there would be points $(\rho, v), (\bar{\rho}, \bar{v})$ such that $u(\rho, v)) = u(\bar{\rho}, \bar{v})$. Since $\frac{\partial u_1}{\partial \rho} \neq 0$, we may assume that $v \neq \bar{v}$. Now using (2.2.26), we have

$$\frac{\rho}{\bar{\rho}} \left\{ \frac{\sigma^2 + c^2}{c^2} \frac{v^2}{c^2 - v^2} + 1 \right\} = \left\{ \frac{\sigma^2 + c^2}{c^2} \frac{\bar{v}^2}{c^2 - \bar{v}^2} + 1 \right\},$$

and

$$\frac{\rho}{\bar{\rho}} \left\{ \frac{v}{c^2 - v^2} \right\} = \frac{\bar{v}}{c^2 - \bar{v}^2}.$$

Eliminating $\frac{\rho}{\bar{\rho}}$ and simplifying gives

$$(v - \bar{v}) \left\{ \frac{\sigma^2}{c^2} v\bar{v} - c^2 \right\} = 0.$$

Since $v \neq \bar{v}$, this implies

$$\frac{\sigma^2}{c^2} v\bar{v} - c^2 = 0,$$

which contradicts the assumptions $|\sigma| < c, |v| < c$ and $|\bar{v}| < c$. Thus the mapping is $1 - 1$. A straightforward calculation shows that

$$det \left(\frac{\partial(u^1, u^2)}{\partial(\rho, v)} \right) = \frac{\rho(\sigma^2 + c^2)}{c^2(c^2 - v^2)^2} \left\{ c^4 - v^2\sigma^2 \right\} > 0. \square$$

It is important to note that the systems (2.1.14) and (2.1.15) are Lorentz invariant. This means that under any Lorentz transformation $(t, x) \rightarrow (\bar{t}, \bar{x})$, one obtains an identical system in the barred coordinates once the velocity states are renamed in terms of the coordinate velocities as measured in the barred coordinate system. Thus, in particular, under Lorentz transformations, $\rho(t, x)$ is a scalar invariant, and thus it takes the same value in the barred and unbarred coordinates that name the same geometric point in the background spacetime manifold. On the other hand, the velocity v is not a scalar, since it is formed from the entries of the vector quantity (u^0, u^1). In this section we will exploit the transformation law for velocities by calculating the shock curves and shock speeds in a frame in which the particle velocity v is zero, and then applying the Lorentz transformation law for velocities to obtain these quantities in an arbitrary frame. The velocity transformation law can be given as follows (c.f. [35]): If in a Lorentz transformation, the barred frame $(\bar{t}, \bar{x})$ moves with velocity μ as measured in the unbarred frame (t, x), and if v denotes the velocity of a particle as measured in the unbarred frame, and $\bar{v}$ the velocity of the same particle as measured in the barred frame, then

$$v = \frac{\mu + \bar{v}}{1 + \frac{\mu\bar{v}}{c^2}}. \tag{2.2.28}$$

Since under Lorentz transformations ρ transforms like a scalar but v does not, it follows that the estimate (2.1.20), which is based on the scalar ρ and *not* the velocity v, expresses a Lorentz invariant property of the weak solutions of (2.1.14), (2.1.15). On the other hand, the estimate (2.1.21) is based on the quantity $\ln\left(\frac{c+v}{c-v}\right)$, which is *not* a Lorentz invariant scalar quantity. Nevertheless, it turns out, (remarkably!), that $Var\{\ln\left(\frac{c+v(\cdot,t)}{c-v(\cdot,t)}\right)\}$ is Lorentz invariant, and thus (2.1.21), as well as (2.1.20), is a Lorentz invariant statement. This is a consequence of the following result.

Proposition 2 *Let $v(x, t)$ be any velocity field which satisfies the velocity transformation law (2.2.28) under Lorentz transformations. Then*

$$Var_x \left\{ \ln\left(\frac{c + v(x, t)}{c - v(x, t)} \right) \right\} = Var_x \left\{ \ln\left(\frac{c + \bar{v}(L(x, t))}{c - \bar{v}(L(x, t))} \right) \right\}, \tag{2.2.29}$$

where L is any Lorentz transformation, $\bar{\mathbf{x}} = L\mathbf{x}$, and v and $\bar{v}$ are related by (2.2.28).

Proof: By (2.2.28), v and $\bar{v}$ are related by the equation

$$v(x,t) = \frac{\mu + \bar{v}(L(x,t))}{1 + \frac{\mu\bar{v}(L(x,t))}{c^2}},$$

where μ is the velocity of the barred frame $\bar{\mathbf{x}}$ as measured in the unbarred frame $\mathbf{x}$. Then for any $x_{i-1} < x_i$, this implies

$$\ln\left\{\frac{c + v(x_i, t)}{c - v(x_i, t)}\right)\} - \ln\left\{\frac{c + v(x_{i-1}, t)}{c - v(x_{i-1}, t)}\right\}$$

$$= \ln\left\{\left(\frac{c + v(x_i, t)}{c - v(x_i, t)}\right)\left(\frac{c + v(x_{i-1}, t)}{c - v(x_{i-1}, t)}\right)^{-1}\right\}$$

$$= \ln\left\{\left(\frac{c + \bar{v}(x_i, t)}{c - \bar{v}(x_i, t)}\right)\left(\frac{c + \bar{v}(x_{i-1}, t)}{c - \bar{v}(x_{i-1}, t)}\right)\right\}$$

$$= \ln\left\{\frac{c + v(x_i, t)}{c - v(x_i, t)}\right\} - \ln\left\{\frac{c + v(x_{i-1}, t)}{c - v(x_{i-1}, t)}\right\},$$

from which (2.2.29) follows.□

2.3 The Wave Speeds

In this section we construct the eigenvalues and Riemann invariants that are associated with a the system of conservation laws (2.1.15).

First recall the three important velocities associated with a system (2.1.15): the particle velocity v, the wave speeds $\lambda_i(\rho, v)$ and the shock speeds $s_i(\rho, v)$, $i = 1, 2$. The wave speeds are the speeds of propagation of the characteristic curves, and for (2.2.24), the λ_i are the eigenvalues of the 2×2 matrix of derivatives $dF \equiv \partial F/\partial u$. Thus $dF \cdot R_i = \lambda_i R_i$, where R_i denotes the i'th right eigenvector of dF. For weak solutions of (2.2.24), discontinuities propogate at the shock speeds s_i which are determined from the Rankine-Hugoniot jump relations (see [17])

$$s[u] = [F]. \tag{2.3.30}$$

Here $[f] \equiv f_L - f_R$ denotes the jump in the function $f(u)$ between the left and right hand states along the curve of discontinuity in the xt-plane. It is not *apriori* clear that a characteristic curve or shock curve $(x(\tau), t(\tau))$, computed in one Lorentz coordinate system will transform to the same spacetime curve when computed in a different Lorentz frame. In the Appendix we will show that both properties are a consequence of the conservation form of the equations. It follows that the derivative $(x'(\tau), t'(\tau))$ transforms like a vector field and that the corresponding speed $x'(\tau)/t'(\tau)$ transforms by the relativistic transformation law for velocities. Thus we can conclude that λ_i and s_i transform according to (2.2.28) under a Lorentz transformation. Since in the system (2.1.15), the flux F is given implicitly as a rather complicated function of u, it is convenient to note that λ_i, R_i, and s_i can all be calculated from the jump relation (2.3.30) alone. For this we need the following well-known theorem due to Lax, [17, 26]:

Theorem 2 *Assume that the system (2.2.24) is strictly hyperbolic; i.e., that $\lambda_1 < \lambda_2$ in the physical domain of u. Then, for a fixed state u_L, the solutions u and s of the Rankine-Hugoniot relation $s(u - u_L) = F(u) - F(u_L)$ can be described (in a neighborhood of u_L) by two families of smooth curves $u = S_i(\varepsilon), S_i(0) = u_L$, with corresponding speeds $s_i(\varepsilon), i = 1, 2$. Moreover, as $\varepsilon \to 0$, we have $s_i(\varepsilon) \to \lambda_i(u_L)$ and $u_i'(\varepsilon) \to R_i$. Here, the parameter ε can be taken to be (Euclidean) arclength along the shock curve S_i in u-space.*

We now use Theorem 2 to obtain the eigenpairs $(\lambda_i, R_i), i = 1, 2$, for system (2.1.15) in the case of a general equation of state of the form $p = p(\rho)$. To start, write system (2.1.15) in the form

$$A_t + B_x = 0,$$

$$B_t + C_x = 0,$$

$$(2.3.31)$$

where

$$A = \frac{(p + \rho c^2)}{c^2} \frac{v^2}{c^2 - v^2} + \rho,$$

$$B = (p + \rho c^2)\frac{v}{c^2 - v^2},$$

$$C = (p + \rho c^2)\frac{v^2}{c^2 - v^2} + p.$$

$$(2.3.32)$$

Then by (2.3.30), for fixed (ρ_L, v_L), the state $(\rho, v) \equiv (\rho_R, v_R)$ lies on a shock curve if and only if

$$[B]^2 = [A][C],$$

$$(2.3.33)$$

where, for example, $[A] \equiv A - A_L$ and $A \equiv A(\rho, v)$ is a function of the unknowns ρ and v along the shock curve. Now assuming that (2.3.33) defines v implicitly as a function of ρ, (this assumption is justified by the construction of the solution itself) differentiate (2.3.33) with respect to ρ and divide by $[B]$ to obtain

$$2B' = \frac{[A]}{[B]}C' + \frac{[C]}{[B]}A',$$

$$(2.3.34)$$

where prime denotes $d/d\rho$. We first obtain a formula for $dv/d\rho$ evaluated at $\rho = \rho_L$. To this end, note that by L'Hospital's rule,

$$\lim_{\rho \to \rho_L} \frac{[A]}{[B]} = \frac{A'}{B'},$$

where the right hand side is evaluated at $\rho = \rho_L$. Thus, at $\rho = \rho_L$, equation (2.3.34) reduces to

$$(B')^2 = A'C'. \qquad (2.3.35)$$

Using (2.3.32) we find

$$A' = \frac{(p' + c^2)}{c^2} e + \frac{(p + \rho c^2)}{c^2} \frac{de}{dv} v' + 1,$$

$$B' = (p' + c^2) \frac{e}{v} + (p + \rho c^2) \frac{(v \frac{de}{dv} - e) v'}{v^2}, \qquad (2.3.36)$$

$$C' = (p' + c^2) e + (p + \rho c^2) \frac{de}{dv} v' + p',$$

where

$$e \equiv \frac{v^2}{c^2 - v^2}, \qquad (2.3.37)$$

$$\frac{de}{dv} = \frac{2c^2 v}{c^2 - v^2}, \qquad (2.3.38)$$

and all terms are evaluated at $\rho = \rho_L$. Substituting (2.3.36) into (2.3.35) and collecting like powers of v' yields

$$0 = (v')^2 \{ (p + \rho c^2)^2 \frac{(v \frac{de}{dv} - e)^2}{v^4} - \frac{(p + \rho c^2)^2}{c^2} (\frac{de}{dv})^2 \}_1 \qquad (2.3.39)$$

$$+ v' \{ \frac{2}{v^3} (p' + c^2)(p + \rho c^2) e (v \frac{de}{dv} - e) - \frac{(p + \rho c^2)}{c^2} e' (p' + 2(p' + c^2) e + 1) \}_2$$

$$+ \{ (p' + c^2)^2 \frac{e^2}{v^2} - \frac{(p' + c^2) e}{c^2} + 1)((p' + c^2) e + p') \}_3.$$

Here we label the brackets so that we can evaluate them separately. A calculation using (2.3.37) and (2.3.38) in (2.3.39) gives

$$\{ \cdot \}_1 = \frac{(p + \rho c^2)^2}{(c^2 - v^2)^2},$$

$$\{ \cdot \}_2 = 0,$$

$$\{ \cdot \}_3 = -p'.$$

Substituting these into (2.3.39) we conclude

$$\frac{v'}{c^2 - v^2} = \overset{+}{_-} \frac{\sqrt{p'}}{p + \rho c^2}. \qquad (2.3.40)$$

We can now solve for the Riemann invariants associated with system (2.1.15). Recall that a Riemann invariant for (2.2.24) is a scalar function $f, \nabla f \neq 0$, which is constant along the integral curves of one of the eigenvector fields of matrix field dF. By Theorem 2, the shock curves $\mathbf{S_i}$ are tangent to the eigenvectors R_i at $\rho = \rho_L$, and thus the Riemann invariants for system (2.1.15) satisfy the differential equations

$$\frac{dv}{d\rho} = \pm \frac{\sqrt{p'}(c^2 - v^2)}{p + \rho c^2},$$

which have the solutions

$$\frac{1}{2} \ln \left(\frac{c+v}{c-v} \right) = \pm c \int \frac{\sqrt{p'(s)}}{p(s) + c^2 s} ds. \qquad (2.3.41)$$

Thus we may define a pair of Riemann invariants r and s for system (2.1.15) as

$$r = \frac{1}{2} \ln \left(\frac{c+v}{c-v} \right) - c \int_1^\rho \frac{\sqrt{p'(s)}}{p(s) + c^2 s} ds, \qquad (2.3.42)$$

$$s = \frac{1}{2} \ln \left(\frac{c+v}{c-v} \right) + c \int_1^\rho \frac{\sqrt{p'(s)}}{p(s) + c^2 s} ds. \qquad (2.3.43)$$

the eigenvalues $\lambda_i(\rho, v)$ of system (2.1.15). By Theorem 2, these are obtained as the limit of s in (2.3.30) as $\rho \to \rho_L$. Solving for s in (2.3.31) we obtain

$$s[A] = [B].$$

Thus, again assuming that $v = v(\rho)$, Theorem 2 implies

$$\begin{aligned}
\lambda_i &= \lim_{\rho \to \rho_L} \frac{[B]}{[A]} \\
&= \lim_{\rho \to \rho_L} c \frac{(p + \rho c^2)e + p}{(p + \rho c^2)e + \rho c^2} \\
&= \lim_{\rho \to \rho_L} c \frac{(p' + c^2)e + (p + \rho c^2)\frac{de}{dv}v' + p'}{(p' + c^2)e + (p + \rho c^2)\frac{de}{dv}v' + c^2}, \qquad (2.3.44)
\end{aligned}$$

where we have applied L'Hospital's rule. But from (2.3.40),

$$\frac{dv}{d\rho} = \pm \frac{\sqrt{p'}}{p + \rho c^2}(c^2 - v^2), \qquad (2.3.45)$$

so substituting (2.3.45) into (2.3.44) and simplifying gives

$$\lambda_1 = \frac{v - \sqrt{p'}}{1 - \frac{v\sqrt{p'}}{c^2}}, \qquad (2.3.46)$$

and

$$\lambda_2 = \frac{v + \sqrt{p'}}{1 + \frac{v\sqrt{p'}}{c^2}}. \qquad (2.3.47)$$

One can now verify directly that r [resp.s] is a 1-[resp. 2-] Riemann invariant for system (2.1.15), by which we mean that r [resp. s] is constant along integral curves of R_2 [resp. R_1].

We now utilize the formulas (2.3.42), (2.3.43), (2.3.46) and (2.3.47) to obtain conditions on p under which the system (2.1.15) is strictly hyperbolic and genuinely nonlinear in the sense of Lax [17]. Recall that a system of conservation laws (2.2.24) is said to be genuinely nonlinear in the i'th characteristic family (λ_i, R_i) if $\nabla_u \lambda_i \cdot R_i \neq 0$ at each point u in state space. (Here, the ∇ denotes the standard Euclidean gradient on state space (ρ, v).) The following theorem gives necessary and sufficient conditions for system (2.1.15), with barotropic equation of state $p = p(\rho)$, to be strictly hyperbolic and genuinely nonlinear.

Theorem 3 *System (2.1.15) is strictly hyperbolic at (ρ, v) if and only if*

$$\sqrt{p'(\rho)} < c. \tag{2.3.48}$$

Moreover, assuming (2.3.48), system (2.1.15) is genuinely nonlinear at each (ρ, v) if and only if

$$p''(\rho) \geq -2 \frac{(c^2 - p')p'}{p + \rho c^2}. \tag{2.3.49}$$

Inequality (2.3.48) is also necessary and sufficient for the sound speeds λ_i to be bounded by c. Moreover, note that the condition for genuine nonlinearity is a geometric condition, being a condition involving only a function of the scalar invariant ρ. Indeed, by Lorentz invariance we know apriori that the condition for genuine nonlinearity could not have involved the state variable v, for if it did, then we reach the absurd conclusion that a Lorentz change of frame could change the wave structure of solutions.

Proof: When $\sqrt{p'(\rho)} = c$, $\lambda_i = {}^+_- c$, so (2.3.48) is required for $|\lambda| \leq c$. It is straightforward to verify that when $\sqrt{p'(\rho)} < c$, $\lambda_1 < \lambda_2$ holds. To verify (2.3.49), note that r and s are constant on integral curves of R_2 and R_1, respectively, so that we may take R_1 and R_2 parallel to $\frac{\partial}{\partial r} = \frac{\partial \rho}{\partial r} \frac{\partial}{\partial \rho} + \frac{\partial v}{\partial r} \frac{\partial}{\partial v}$ and $\frac{\partial}{\partial s} = \frac{\partial \rho}{\partial s} \frac{\partial}{\partial \rho} + \frac{\partial v}{\partial s} \frac{\partial}{\partial v}$, respectively. Differentiating $r + s = \ln\left(\frac{c+v}{c-v}\right)$ and $s - r = c \int_1^\rho \frac{\sqrt{p'(s)}}{p(s)+c^2 s} ds$, we can solve for the partials of ρ and v with respect to r and s and obtain, (up to positive scale factor),

$$R_1 \equiv \left(\frac{-c}{c^2 - v^2}, \frac{c\sqrt{p'}}{p + \rho c^2} \right)^t, \tag{2.3.50}$$

and

$$R_2 \equiv \left(\frac{c}{c^2 - v^2}, \frac{c\sqrt{p'}}{p + \rho c^2} \right)^t. \tag{2.3.51}$$

Using these it is straightforward to verify that

$$\nabla\lambda_1 \cdot R_1 = \frac{c^2}{(\sqrt{p'}v - c^2)^2}\left\{\frac{(c^2 - p')\sqrt{p'}}{p + \rho c^2} + \frac{p''}{2\sqrt{p'}}\right\} > 0 \qquad (2.3.52)$$

and

$$\nabla\lambda_2 \cdot R_2 = \frac{c^2}{(\sqrt{p'}v + c^2)^2}\left\{\frac{(c^2 - p')\sqrt{p'}}{p + \rho c^2} + \frac{p''}{2\sqrt{p'}}\right\} > 0, \qquad (2.3.53)$$

if and only if (2.3.49) holds.□

Note that from the formulas (2.3.50) and (2.3.51) it follows that ρ decreases [resp. increases] and v increases along the integral curves of R_1 [resp. R_2] in the direction of increasing λ_1 [resp. λ_2]. Thus, the integral curves of R_i give the oriented rarefaction curves for system (2.1.15), [17, 26].

2.4 The Shock Curves

In this section we restrict to the case $p = \sigma^2\rho$ and obtain global parameterizations of the shock curves $\mathbf{S_i}$ and shock speeds s_i which are valid for arbitrary ρ_L and v_L. (Here we let $\mathbf{S_i}$ denote the portion of $\mathbf{S_i}(\varepsilon)$, defined in Theorem 2, which corresponds to shocks satisfying the Lax condition. Upon normalizing R_i by $\nabla\lambda_i \cdot R_i > 0$, we can assume without loss of generality that $\mathbf{S_i}$ is given locally by $\varepsilon \leq 0$, [17, 26].) We use our global parameterization to study the geometry of the shock curves in Riemann invariant coordinates (r, s).

Lemma 1 *Assume that (ρ_L, v_L) and $(\rho, v) \equiv (\rho_R, v_R)$ satisfy the jump conditions (2.3.30) for system (2.1.15) with equation of state (2.1.7). Then the following relations hold:*

$$\frac{\rho}{\rho_L} = 1 + \beta\left\{1 \mp \sqrt{1 + \frac{2}{\beta}}\right\} \equiv f_{\stackrel{-}{+}}(\beta), \qquad (2.4.54)$$

where

$$\beta \equiv \beta(v, v_L) = \frac{(\sigma^2 + c^2)^2}{2\sigma^2}\frac{(v - v_L)^2}{(c^2 - v^2)(c^2 - v_L^2)}. \qquad (2.4.55)$$

The shock curve S_1 is given by (2.4.54) and parameterized by $\rho \geq \rho_L$ when we take a plus sign in (2.4.54), and S_2 is given by (2.4.54) and parameterized by $\rho \leq \rho_L$ when we take the minus sign.

Proof: For system (2.1.15), the Rankine-Hugoniot jump conditions (2.3.30) give

$$s[\rho\frac{(\sigma^2 + c^2)}{c^2}\frac{v^2}{c^2 - v^2} + \rho] = [\rho(\sigma^2 + c^2)\frac{v}{c^2 - v^2}],$$

and

$$s[\rho(\sigma^2 + c^2)\frac{v}{c^2 - v^2}] = [\rho(\sigma^2 + c^2)\frac{v^2}{c^2 - v^2} + \rho\sigma^2].$$

Eliminating the shock speed s and recalling $u^2 = v^2/(c^2 - v^2)$ gives

$$\left\{\frac{c\rho v}{c^2 - v^2} - \frac{c\rho_L v_L}{c^2 - v_L^2}\right\}^2 = \left\{\frac{\rho v^2}{c^2 - v^2} + \frac{\rho c^2}{\sigma^2 + c^2} - \frac{\rho_L v_L^2}{c^2 - v_L^2} - \frac{\rho_L c^2}{\sigma^2 + c^2}\right\}$$
$$\times \left\{\frac{\rho v^2}{c^2 - v^2} + \frac{\rho \sigma^2}{\sigma^2 + c^2} - \frac{\rho_L v_L^2}{c^2 - v_L^2} - \frac{\rho_L \sigma^2}{\sigma^2 + c^2}\right\}.$$

Now multiplying out gives

$$0 = \left(\frac{\rho - \rho_0}{\rho_0}\right)^2 \left\{u^4 + u^2 - \frac{c^2 u^4}{v^2} + \frac{\sigma^2 c^2}{(\sigma^2 + c^2)^2}\right\}_1$$
$$+ \left(\frac{\rho - \rho_0}{\rho_0}\right)\left\{2u^2[u^2] + [u^2] - \frac{2c^2 u^2}{v}\left[\frac{u^2}{v}\right]\right\}_2 \qquad (2.4.56)$$
$$+ \left\{[u^2]^2 - c^2\left[\frac{u^2}{v}\right]^2\right\}_3.$$

Simplifying, we obtain

$$\{\cdot\}_1 = \frac{\sigma^2 c^2}{(\sigma^2 + c^2)^2},$$

$$\{\cdot\}_2 = -\frac{c^2(v - v_L)^2}{(c^2 - v^2)(c^2 - v_L^2)},$$

and

$$\{\cdot\}_3 = -\frac{c^2(v - v_L)^2}{(c^2 - v^2)(c^2 - v_L^2)},$$

so that (2.4.56) becomes

$$0 = \left(\frac{\rho - \rho_0}{\rho_0}\right)^2 \left\{\frac{\sigma^2 c^2}{(\sigma^2 + c^2)^2}\right\}_1$$
$$- \left(\frac{\rho - \rho_0}{\rho_0}\right)\left\{\frac{c^2(v - v_L)^2}{(c^2 - v^2)(c^2 - v_L^2)}\right\}_2 \qquad (2.4.57)$$
$$- \left\{\frac{c^2(v - v_L)^2}{(c^2 - v^2)(c^2 - v_L^2)}\right\}_3.$$

Now substituting β (given in (2.4.55)) in (2.4.57) and solving for $(\rho - \rho_0)/\rho_0$ yields (2.4.54).

From Theorem 2 together with (2.3.50) and (2.3.51), it follows that for the shock curve $\mathbf{S}_i$, which is tangent to R_i at $\rho = \rho_L$, we must take a plus sign for $i = 1$ and a minus sign when $i = 2$. Moreover, by Theorem 2 together with (2.3.52) and (2.3.53), the S_1 is parameterized when $\rho \geq \rho_L$, and S_2 is parameterized when $\rho \leq \rho_L$. $\Box$

The general theory of conservation laws only guarantees that the "Lax shock conditions" hold on $\mathbf{S_i}$ near $\rho = \rho_L$, [26]. We will show that these actually hold all along the shock curves $\mathbf{S_i}$. We state this as part of the following lemma.

Lemma 2 *Let $p = \sigma^2 \rho$. Then the the shock speed s_i is monotone all along the shock curve $\mathbf{S_i}$ of system (2.1.15), $i = 1, 2$, and moreover, the following inequalities (Lax shock conditions) hold at each state $u_R \neq u_L$ on the shock curve $\mathbf{S_i}$:*

$$\lambda_i(\rho_L, v_L) < s_i < \lambda_i(\rho_R, v_R). \tag{2.4.58}$$

Proof: First, since the shock speeds s_i and wave speeds λ_i transform by the velocity transformation law (2.2.28) (see appendix), it suffices to verify (2.4.58) in the case $v_L = 0$. In this case we set $\rho_L = \rho_0$. Note that by Lemma 1, $v \leq v_L = 0$ all along both S_1 and S_2. We first obtain s_i and λ_i along $\mathbf{S_i}$ in terms of the parameter β defined in (2.4.55). So, assuming $p = \sigma^2 \rho$, the jump conditions (2.3.30) applied to system (2.1.15) directly give

$$s^2 = \left\{ \left(\frac{z}{z-1}\right) \frac{c^2 v^2}{c^2 - v^2} + \frac{c^2 \sigma^2}{c^2 + \sigma^2} \right\} \Big/ \left\{ \left(\frac{z}{z-1}\right) \frac{v^2}{c^2 - v^2} + \frac{c^2}{c^2 + \sigma^2} \right\}, \tag{2.4.59}$$

where we have set

$$z = \frac{\rho}{\rho_0}. \tag{2.4.60}$$

Now it follows from (2.4.54) that

$$\frac{z}{z-1} = \frac{\beta \mp \sqrt{\beta^2 + 2\beta}}{\beta \mp \sqrt{\beta^2 + 2\beta}},$$

and from (2.4.55) that

$$v^2 = c^2 K \frac{\beta}{1 + K\beta},$$

where we set

$$K = \frac{2\sigma^2 c^2}{(\sigma^2 + c^2)^2}. \tag{2.4.61}$$

Using these in (2.4.59) gives (after simplification)

$$s^2 = c^2 \frac{f_{\mp}(\beta) + \frac{\sigma^2}{c^2}}{f_{\mp}(\beta) + \frac{c^2}{\sigma^2}}, \tag{2.4.62}$$

where

$$f_{\stackrel{-}{+}}(\beta) = 1 + \beta \stackrel{-}{+} \sqrt{\beta^2 + 2\beta}. \tag{2.4.63}$$

Note that the plus/minus sign in the above formulas agrees with the plus/minus sign in (2.4.54). Thus, S_1, is parameterized by $0 \le \beta < \infty$ (starting at (ρ_L, v_L)) when we choose the plus sign in (2.4.63) and S_2 is parameterized by $0 \le \beta < \infty$ when we choose the minus sign in (2.4.63). Specifically, on S_1,

$$s_1(\beta) = -c\sqrt{\frac{1 + \beta + \sqrt{\beta^2 + 2\beta} + \frac{\sigma^2}{c^2}}{1 + \beta + \sqrt{\beta^2 + 2\beta} + \frac{c^2}{\sigma^2}}}; \tag{2.4.64}$$

and on S_2,

$$s_2(\beta) = -c\sqrt{\frac{1 + \beta - \sqrt{\beta^2 + 2\beta} + \frac{\sigma^2}{c^2}}{1 + \beta - \sqrt{\beta^2 + 2\beta} + \frac{c^2}{\sigma^2}}}. \tag{2.4.65}$$

We now obtain a corresponding parameterization of λ_i with repect to β along $\mathbf{S_i}$. In the case $p = \sigma^2 \rho$, the formulas (2.3.46) and (2.3.47) give

$$\lambda_1 = \frac{v - \sigma}{1 - \frac{\sigma v}{c^2}}, \tag{2.4.66}$$

and

$$\lambda_2 = \frac{v + \sigma}{1 + \frac{\sigma v}{c^2}}. \tag{2.4.67}$$

Thus

$$\lambda_1 = c^2 \frac{\sigma + \sqrt{c^2 K \frac{\beta}{1+K\beta}}}{-\sigma\sqrt{c^2 K \frac{\beta}{1+K\beta}} - c^2}, \tag{2.4.68}$$

and

$$\lambda_2 = c^2 \frac{\sigma - \sqrt{c^2 K \frac{\beta}{1+K\beta}}}{-\sigma\sqrt{c^2 K \frac{\beta}{1+K\beta}} + c^2}, \tag{2.4.69}$$

where we have used $v \le 0$.

Now differentiating (2.4.62) gives (for $\beta \ne 0, \sigma < c$,)

$$\frac{ds_i^2}{d\beta} = c^2 \frac{\left(\frac{c^2}{\sigma^2} - \frac{\sigma^2}{c^2}\right) f'_+(\beta)}{\left(f_{\stackrel{-}{+}}(\beta) + \frac{c^2}{\sigma^2}\right)^2} \ne 0, \tag{2.4.70}$$

from which the monotonicity of shock speeds along the shock curves easily follows. The inequality $s_i < \lambda_i(\rho_L, 0)$ thus holds. It remains to show that $\lambda_i < s_i$ along the shock curve $\mathbf{S_i}$. We do the case $i = 1$. To this end, let $x \equiv \sigma/c$ and set $z \equiv \rho/\rho_0 = f_-(\beta)$. Substituting into (2.4.64) and (2.4.66) yields

$$s_1^2 - \lambda_1^2 = c^2 \frac{\left(x - \sqrt{\frac{\sqrt{K}(z-1)}{\sqrt{2z+K(1-z)^2}}}\right)^2}{x^2(z + \frac{1}{x^2})}$$

$$\cdot \left\{ \frac{x^2(z+x^2)}{\left(x - \sqrt{\frac{\sqrt{K}(z-1)}{\sqrt{2z+K(1-z)^2}}}\right)^2} - \frac{z + \frac{1}{x^2}}{\left(\frac{1}{x} - \sqrt{\frac{\sqrt{K}(z-1)}{\sqrt{2z+K(1-z)^2}}}\right)^2} \right\}$$

$$= \frac{c^2 x^{-2}(z + \frac{1}{x^2})^{-1} \left(x - \sqrt{\frac{\sqrt{K}(z-1)}{\sqrt{2z+K(1-z)^2}}}\right)^2}{\left(x\sqrt{2z + K(1-z)^2} - \sqrt{K}(z-1)\right)^2}$$

$$\cdot \left\{ \frac{\sqrt{2z + K(1-z)^2}}{\left(\frac{1}{x}\sqrt{2z + K(1-z)^2} - \sqrt{K}(z-1)\right)^2} \right\} \{\cdot\}_* ,$$

$$(2.4.71)$$

where

$$\{\cdot\}_* = \left(x^2(z + x^2)\right)\left(\frac{1}{x}\sqrt{2z + K(1-z)^2} - \sqrt{K}(z-1)\right)^2$$
$$- \left(z + \frac{1}{x^2}\right)\left(x\sqrt{2z + K(1-z)^2} - \sqrt{K}(z-1)\right)^2 .$$

Thus it suffices to show that $\{\cdot\}_* < 0$. A calculation using the identities

$$\frac{x^4 + x^2 + 1}{x^2} = \frac{2}{K} - 1,$$

and

$$\frac{x^2 + 1}{x} = \frac{2}{K},$$

leads to

$$\{\cdot\}_* = (1 - x^2)(1 - z)\left\{\frac{2}{K} + \frac{2^{3/2}}{K}\sqrt{2z + K(1-z)^2}\right\}. \qquad (2.4.72)$$

But by (2.4.72) it follows that $\{\cdot\} < 0$ for $z > 1$, and thus by (2.4.71) we must have $s_1^2 - \lambda_1^2 < 0$. Since both $s_1 < 0$ and $\lambda_1 < 0$ along S_1 when $v_L = 0$, it follows that $\lambda_1 < s_1$ all along S_1, thus finishing the proof of Lemma 2.$\square$

2.5 Geometry of the Shock Curves

In this section we study the special geometry of the shock curves in the plane of Riemann invariants for system (2.1.15), the case $p = \sigma^2 \rho$. In this case the

the shock curves are given by (2.4.54) and (2.4.55), and using (2.3.42) and (2.3.43), the Riemann invariants r and s are given in this case by, (see Figure 1),

$$r = \frac{1}{2}\ln\left(\frac{c+v}{c-v}\right) - \frac{K_0}{2}\ln(\rho),$$ (2.5.73)

$$s = \frac{1}{2}\ln\left(\frac{c+v}{c-v}\right) + \frac{K_0}{2}\ln(\rho),$$ (2.5.74)

where

$$K_0 = \sqrt{\frac{K}{2}} = \frac{\sigma c}{c^2 + \sigma^2},$$

and K is defined in (2.4.61).

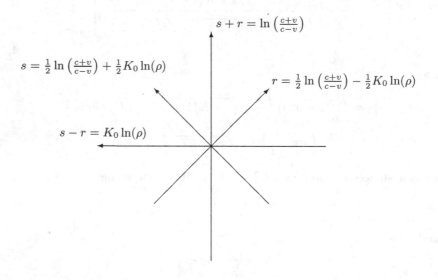

The plane of Riemann invariants.

Figure 1

Our main result of this section is that the i-shock curves are independent of the base point (ρ_L, v_L) in the sense that, when graphed in the rs-plane, all i-shock curves are rigid translations of one another; and moreover, the 1-shock curve based at a given point is the reflection of the 2-shock curve based at the same point about an appropriate axis of rotation. On an algebraic level, this happens because, along a shock curve, $\rho\rho_L^{-1}$ and $\left\{\frac{c+v}{c-v}\right\}\left\{\frac{c+v_L}{c-v_L}\right\}^{-1}$ in the definitions of r and s turn out to be functions of the parameter β alone, and the functions that give $\rho\rho_L^{-1}$ as a function of β are reciprocal on the 1-and 2-shock curves, respectively. We begin with the following lemma, which gives a parameterization of the i-shock curves for system (2.1.15) in the rs-plane:

Lemma 3 *Let* $r \equiv r(\rho, v), s \equiv s(\rho, v), \Delta r \equiv r(\rho, v) - r(\rho_L, v_L)$ *and* $\Delta s \equiv s(\rho, v) - s(\rho_L, v_L)$ *where we let* $(\rho, v) \equiv (\rho_R, v_R)$. *Then the 1-shock curve* S_1 *for system (2.1.15) based at* (r_L, s_L) *is given by the following parameterization with respect to the parameter* β, $0 \le \beta < \infty$:

$$\Delta r = -\frac{1}{2}\ln\{f_+(2K\beta)\} - \sqrt{\frac{K}{2}}\ln\{f_+(\beta)\}, \qquad (2.5.75)$$

$$\Delta s = -\frac{1}{2}\ln\{f_+(2K\beta)\} + \sqrt{\frac{K}{2}}\ln\{f_+(\beta)\}; \qquad (2.5.76)$$

and the 2-shock curve S_2 *based at* (r_L, s_L) *is given for* $0 \le \beta < \infty$, *by*

$$\Delta r = -\frac{1}{2}\ln\{f_+(2K\beta)\} - \sqrt{\frac{K}{2}}\ln\{f_-(\beta)\}, \qquad (2.5.77)$$

$$\Delta s = -\frac{1}{2}\ln\{f_+(2K\beta)\} + \sqrt{\frac{K}{2}}\ln\{f_-(\beta)\}. \qquad (2.5.78)$$

Proof: For convenience, define

$$w \equiv \frac{c - v}{c + v}, \qquad (2.5.79)$$

so that

$$\frac{v}{c} = \frac{1 - w}{1 + w},$$

and by (2.5.73) and (2.5.74),

$$r + s = -\ln(w).$$

Then by the definition of β in (2.4.55) we have

$$K\beta(v, v_L) = \frac{(\frac{v}{c} - \frac{v_L}{c})^2}{(1 - \frac{v^2}{c^2})(1 - \frac{v_L^2}{c^2})} \qquad (2.5.80)$$

$$= \left(\frac{1 - w}{1 + w} - \frac{1 - w_L}{1 + w_L}\right)^2 \frac{(1 + w_L)^2}{4w} \frac{(1 + w_L)^2}{4w_L}$$

$$= \frac{1}{4}\left\{\sqrt{\frac{w}{w_L}} - \sqrt{\frac{w_L}{w}}\right\}^2,$$

which we can rewrite as

$$K\beta = \frac{1}{4}\left\{ exp\{\frac{r-r_L}{2} + \frac{s-s_L}{2}\} - exp\{-\frac{r-r_L}{2} - \frac{s-s_L}{2}\}\right\}^2$$

$$= sinh^2\left(\frac{\Delta r}{2} + \frac{\Delta r}{2}\right). \tag{2.5.81}$$

Now solving for w/w_L in (2.5.80) gives

$$\frac{w}{w_L} = 1 + 2K\beta\left(1\underset{+}{\overset{-}{}}\sqrt{1 + \frac{2}{K\beta}}\right) \equiv f_{\underset{+}{-}}(2K\beta). \tag{2.5.82}$$

Note that w is a monotone decreasing function of v, and v decreases along $\mathbf{S_i}$, $i = 1, 2$. Thus $w/w_l \geq 1$ holds along $\mathbf{S_i}$, so that we must choose the plus sign in (2.5.82) on both i-shock curves. On the other hand, by (2.4.54), along the shock curves we have

$$\frac{\rho}{\rho_L} = f_{\underset{+}{-}}(\beta), \tag{2.5.83}$$

where we take the plus sign when $i = 1$ and the minus sign when $i = 2$. Therefore, substituting (2.5.82) and (2.5.83) into (2.5.73) and (2.5.74), and choosing the appropriate plus and minus signs, gives (2.5.75)-(2.5.78). This completes the proof of Lemma 3.□

What is interesting about (2.5.75)-(2.5.78) is that the differences Δr and Δs along a shock curve depend only on the parameter β, and thus the geometric shape of the shock curves in the rs-plane is independent of the base point (r_L, s_L). This immediately implies that an i-shock curve based at one point in the rs-plane can be mapped by a rigid translation onto the i-shock curve based at any other point.

Lemma 4 *The 2-shock curve based at an arbitrary point (r_L, s_L) is the reflection in the rs-plane of the 1-shock curve based at the same point, where the axis of reflection is the line passing through (r_L, s_L), parallel to the line $r = s$.*

Proof: This follows immediatly from (2.5.75)-(2.5.78) because using (2.4.63) we have

$$f_-(\beta)f_+(\beta) = \left(1 + \beta - \beta\sqrt{1 + \frac{2}{\beta}}\right)\left(1 + \beta + \beta\sqrt{1 + \frac{2}{\beta}}\right) = 1. \tag{2.5.84}$$

The following lemma gives further important geometric properties of the shock curves which we shall need.

Lemma 5 *The shock curves S_i given in (2.5.75)-(2.5.78) define convex curves in the rs-plane, and moreover,*

$$0 \leq \frac{ds}{dr} < \frac{\sqrt{2K}-1}{-\sqrt{2K}-1} < 1 \tag{2.5.85}$$

all along a 1-shock curve S_1, and

$$0 \leq \frac{dr}{ds} < \frac{\sqrt{2K} - 1}{-\sqrt{2K} - 1} < 1, \tag{2.5.86}$$

all along a 2-shock curve S_2.

Proof: By symmetry, it suffices to do the case $i = 1$. Differentiating (2.5.75) with respect to β gives

$$\sqrt{\frac{2}{K}} \frac{d\Delta r}{d\beta} = -\frac{1}{\sqrt{2\beta + \beta^2}} - \frac{1}{\sqrt{2\beta + 2K\beta^2}},$$

and differentiating (2.5.76) with respect to β gives

$$\sqrt{\frac{2}{K}} \frac{d\Delta s}{d\beta} = \frac{1}{\sqrt{2\beta + \beta^2}} - \frac{1}{\sqrt{2\beta + 2K\beta^2}},$$

so that

$$\frac{d\Delta s}{d\Delta r} = \frac{ds}{dr} = -\frac{\sqrt{2\beta + 2K\beta^2} - \sqrt{2\beta + \beta^2}}{-\sqrt{2\beta + 2K\beta^2} - \sqrt{2\beta + \beta^2}} < 1,$$

because $2K \leq 1$. To verify the convexity of $\mathbf{S_1}$, we differentiate with respect to β, and simplify to get

$$\frac{d}{d\beta} \frac{ds}{dr} = \frac{2(1 - 2K)}{\beta^2} \left\{ \sqrt{\frac{2}{\beta} + 2K} + \sqrt{\frac{2}{\beta} + 1} \right\}^{-2} \left\{ \sqrt{\frac{2}{\beta} + 2K} \sqrt{\frac{2}{\beta} + 1} \right\}^{-1} \geq 0. \tag{2.5.87}$$

Now (2.5.85) follows from the convexity of the shock curves together with the inequality

$$\lim_{\beta \to \infty} \frac{ds}{dr} = \frac{\sqrt{2K} - 1}{-\sqrt{2K} - 1} \leq 1. \square$$

A graph of the shock curves S_i in the (r, s)-plane is given in Figure 2.

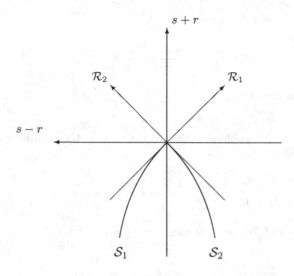

The reflection property of shock curves.

Figure 2

As a final comment in this section, we note that we can obtain a corresponding parameterization of the shock curves for system (2.1.17) by taking the limit $c \to \infty$. To see this, note that taking the limit $c \to \infty$ in (2.5.73) and (2.5.74) we obtain

$$\bar{r} \equiv cr \to v - \sigma \ln(\rho),$$

and

$$\bar{s} \equiv cs \to v + \sigma \ln(\rho),$$

the right hand sides being the Riemann invariants for system (2.1.17). Moreover, under this limit, (2.4.54) remains unchanged, and by (2.4.55),

$$\beta \to \frac{(v - v_L)^2}{2\sigma^2}.$$

Thus multiplying by c and taking the limit $c \to \infty$, the formulas (2.5.75)-(2.5.78) reduce to

$$\Delta \bar{r} = -\frac{\sigma}{\sqrt{2}}\beta - \sigma \ln\{f_+(\beta)\},$$

$$\Delta \bar{s} = -\frac{\sigma}{\sqrt{2}}\beta + \sigma \ln\{f_+(\beta)\};$$

and

$$\Delta \bar{r} = -\frac{\sigma}{\sqrt{2}}\beta - \sigma \ln\{f_-(\beta)\},$$

$$\Delta \bar{s} = -\frac{\sigma}{\sqrt{2}}\beta + \sigma \ln\{f_-(\beta)\},$$

respectively, and this gives parameterizations of the shock curves for system (2.1.17) that are equivalent to the formulas obtained by Nishida in [22].

2.6 The Riemann Problem

In this section we discuss the solution of the Riemann problem for system (2.1.15). In the next section we shall exploit special properties of the solution of the Riemann problem for this system to construct global weak solutions of the general initial value problem by means of the Glimm difference scheme, and to obtain the estimates (2.1.20), (2.1.21) and (2.1.22) .

The Riemann problem is the initial value problem when the initial data $u_0(x) \equiv u(\rho_0(x), v_0(x))$ consists of a pair of constant states $u_L \equiv u(\rho_L, v_L)$ and $u_R \equiv u(\rho_R, v_R)$ separated by a jump discontinuity at $x = 0$,

$$u_0(x) = \begin{cases} u_L \ if \ x < 0, \\ u_R \ if \ x > 0. \end{cases} \tag{2.6.88}$$

Note that, in view of Proposition 1, the conserved quantities u_L and u_R of system (2.1.14) are uniquely determined by (ρ_L, v_L) and (ρ_R, v_R). When u_R is sufficiently close to u_L, the existence and uniqueness of the solution of the Riemann problem for system (2.1.15) in the class of elementary waves follows by a general theorem of Lax which applies to any sytem of conservation laws which is strictly hyperbolic ($\lambda_1 < \lambda_2$) and genuinely nonlinear in each characteristic field. (See Theorem 3, and [17, 26].) We verify that for system (2.1.15), with $p = \sigma^2 \rho$, the solution of the Riemann problem (2.6.88) can be (uniquely) constructed for all u_L and u_R provided that

$$\rho_L > 0, \quad \rho_R > 0, \tag{2.6.89}$$

and

$$-c < v_L < c, \quad -c < v_R < c. \tag{2.6.90}$$

To this end, fix u_L and let $u \equiv u_R$ be variable. Let $\mathbf{R_i} \equiv \mathbf{R_i}(u_L)$ denote the i-rarefaction curve and $\mathbf{S_i} \equiv \mathbf{S_i}(u_L)$ the i-shock curve associated with the state u_L, [26]. The i-rarefaction curve $\mathbf{R_i}$ at u_L is defined to be the segment of the integral curve of the eigenvector R_i which starts at u_L, and continues in the direction of increasing λ_i. Since the Riemann invariants r and s, defined in (2.5.73) and (2.5.74), are constant on the 2- and 1-integral curves, respectively, it follows from genuine nonlinearity and Theorem 3 that

$$\mathbf{R_1}(u_L) = \{u : s(u) = s(u_L) \ and \ r(u) \geq r(u_L)\},$$

and
$$\mathbf{R_2}(u_L) = \{u : r(u) = r(u_L) \ and \ s(u) \geq s(u_L)\}.$$

Note that for each state $u_R \in \mathbf{R_i}(u_L)$, there is a rarefaction wave solution that solves the Riemann problem (2.6.88). Indeed, this is constructed by letting each state u on R_i between u_L and u_R propogate with speed $\lambda_i(u)$.

The i-shock curve $\mathbf{S_i}$ is given in (2.5.75)-(2.5.78) of Lemma 3. If $u_R \in \mathbf{S_i}(\mathbf{u_L})$, then the Riemann problem (2.6.88) is solved (in the weak sense) by a shock wave of speed s_i (given by (2.4.64) and (2.4.65)), and the shock satisfies the Lax admissibility condition, (2.4.58). By Lax's theorem, the curve $\mathbf{S_i}$ makes C^2 tangency with the i'th integral curve at u_L, and thus it follows from (2.5.73), (2.5.74) and (2.5.75)-(2.5.78) that the composite wave curve $\mathbf{T_i}$ defined by $\mathbf{T_i} \equiv \mathbf{S_i} \bigcup \mathbf{R_i}$ is a C^2 curve for each $i = 1, 2$. Let $\mathbf{T_i}(\varepsilon)$ denote the Euclidean arclength parameterization of the i-wave curve $\mathbf{T_i}$, with ε increasing with increasing λ_i ($\varepsilon > 0$ on $\mathbf{R_i}$, $\varepsilon < 0$ on $\mathbf{S_i}$). It follows from (2.5.73), (2.5.74) and (2.5.75)-(2.5.78) that $\ln \rho$ decreases monotonically from $+\infty$ to $-\infty$ along $\mathbf{T_1}$, and increases monotonely from $-\infty$ to $+\infty$ along $\mathbf{T_2}$; furthermore, v increases monotonely from $-c$ to $+c$ along both $\mathbf{T_1}$ and $\mathbf{T_2}$. The wave curves $\mathbf{T_i} \equiv \mathbf{S_i} \bigcup \mathbf{R_i}$ are sketched in Figure 2.

To solve the Riemann problem, consider the wave curves $\mathbf{T_2}(u_M)$ for $u_M \in \mathbf{T_1}(u_L)$. It is easily verified that any two such curves $\mathbf{T_2}(u_M)$ and $\mathbf{T_2}(u'_M), u_M, u'_M \in \mathbf{T_1}(u_L)$, are nonintersecting, and that the set of all such curves covers the entire region $\rho > 0, -c < v < c$ in the ρv-plane in a $1 - 1$ fashion. In particular we use the fact (see Lemma 5) that $|ds/dr| < 1$ and $|dr/ds| < 1$ all along the 1-shock and 2-shock curves, respectively. Now for given states u_L and u_R, let $u_M \in \mathbf{T_1}(u_L)$ denote the unique intermediate state such that $u_R \in \mathbf{T_2}(u_M)$. Then the unique solution of the Riemann problem in the class of elementary waves is given by a 1-wave connecting u_L to u_M, followed by a 2-wave connecting u_M to u_R. It remains only to verify that the 2-wave speed is always greater that the 1-wave speed in this construction. But this follows directly from (2.4.58) of Lemma 2; i.e., if $u_M \in \mathbf{S_1}(u_L)$, and $u_R \in \mathbf{S_2}(u_M)$ then by (2.4.58),

$$s_1 \leq \lambda_1(u_M) < \lambda_2(u_M) \leq s_2.$$

We state these results as a theorem:

Theorem 4 *There exists a solution of the Riemann problem for system (2.1.15) in the case of an equation of state of the form $p = \sigma^2 \rho$, $0 < \sigma < c$, so long as u_L and u_R satisfy (2.6.89) and (2.6.90). Moreover, the solution is given by a 1-wave followed by a 2-wave, satisfies $\rho > 0$, and all speeds are bounded by c. The solution is unique in the class of rarefaction waves and admissible shock waves.*

We note that in the case of an isentropic equation of state, $(p(\rho) = \sigma^2 \rho^\gamma, \gamma > 1)$, the sound speed exceeds the speed of light for sufficiently large ρ, except in the limiting case $\gamma = 1$. Thus, it is only in the limiting case $\gamma = 1$

of the isentropic gas model that a big data global existence theorem for the Riemann problem (such as Theorem 3 above) is possible without modifying the equation of state at large densities.

2.7 The Initial Value Problem

In this section we use the Glimm difference scheme to prove Theorem 1. We begin with a short discussion of this method.

Glimm's method is a procedure for obtaining solutions of the general initial value problem by constructing a convergent sequence of approximate solutions, the approximation scheme being based on the solution of the Riemann problem. The scheme consists of approximating the solution at a fixed time level by piecewise constant states, so that one can solve the resulting Riemann problems thereby obtaining a sequence of elementary waves at that time level, the goal being to estimate the growth in the amplitude of these elementary waves as the waves interact during the time evolution of the solution. Glimm's method provides a scheme by which Riemann problems are *re-posed* at a subsequent time level according to a random choice of the state appearing in the waves of the previous time level. This has the advantage that waves at the subsequent time level are determined through the interaction of waves at the prior time level, and by this scheme, estimates on the amplitude changes in waves during interactions can then be used to estimate the growth of a solution in general. The natural measure of the amplitude, or *strength* of a wave γ, is the magnitude of the jump $|u_R - u_L| \equiv |\gamma|$. Thus, the total wave strength present in an approximate solution at time $t > 0$ is given by

$$\sum_i |\gamma_i|, \qquad (2.7.91)$$

where the sum is over all waves present in the approximate solution at time t. The sum in (2.7.91) is equivalent to the total variation norm of the approximate solution at time $t > 0$. The total variation of the waves will in general increase due to interactions because of the nonlinearity of the equations. Glimm showed that for a strictly hyperbolic, genuinely nonlinear (or linearly degenerate [26]) system, if the total initial strength of waves in an approximate solution is sufficiently small ($\sum_i |\gamma_i| << 1$) at time $t = 0$, then the total strength of waves at time $t > 0$ is bounded by a constant times the initial strength. His method is to define a nonlocal functional Q, quadratic in wave strengths, which has the property that it decreases when waves interact, and moreover this decrease dominates the increase in total wave strength, when the initial wave strength is sufficiently small. This leads to the following theorem:

Theorem 5 (Glimm, [10]) *Consider the initial value problem (2.2.24), (2.2.25) for a strictly hyperbolic, genuinely nonlinear system of conservation* *laws defined in a neighborhood of a state u_*. Then there exist constants $0 < V << 1$, $C > 0$, and a neighborhood $\mathbf{u}$ of u_* such that, if the initial data u_0 lies in $\mathbf{u}$, and*

$$Var\{u_0(\cdot)\} < V, \tag{2.7.92}$$

then there exists a global weak solution $u(x,t)$ of (2.2.24), (2.2.25) obtained by Glimm's method, and this solution satisfies

$$Var\{u(\cdot,t\} \leq CVar\{u_0(\cdot)\}. \tag{2.7.93}$$

Glimm's method of analysis, based on the quadratic potential functional, is the only method by which a time independent bound on wave strengths has been rigorously proven for general systems of hyperbolic conservation laws. In this section we show that system (2.1.15) is better than the general case because it has the very special property that when $p = \sigma^2 \rho$, the total variation of $\ln(\rho)$ is nonincreasing (in time) when elementary waves interact. Thus, $Var\{\ln(\rho)\}$ plays the role of an energy function, and one can use this in place of the nonlocal functional Q in Glimm's method. This fact enables us to prove Theorem 1, a *large data* existence theorem, in this special case. (This idea is due to Nishida [22].)

We now define the Glimm difference scheme for system (2.1.15) in detail, and prove Theorem 1. Let Δx denote a mesh length in x and Δt a mesh length in t, and let $x_i \equiv i\Delta x$ and $t_j \equiv j\Delta t$ denote the mesh points in an approximate solution. Let $u_0(x) \equiv u(\rho_0(x), v_0(x))$ denote initial data for system (2.1.15) satisfying $\rho_0(x) > 0, -c < v_0(x) < c$. To define the Glimm scheme approximate solution $u_{\Delta x}(x,t)$, we approximate the initial data by the piecewise constant states $u_{i0} = u_0(x_i+)$. To start the scheme, define

$$u_{\Delta x}(x,0) = u_{i0}, \quad for \quad x_i \leq x < x_{i+1}.$$

Now assume that the approximate solution $u_{\Delta x}$ has been defined for $t \leq t_{j-1}$, and that the solution at time $t = t_{j-1}$ is given by piecewise constant states

$$u_{\Delta x}(x,t_{j-1}) = u_{i,j-1}, \quad for \quad x_i \leq x < x_{i+1}.$$

In order to complete the definition of $u_{\Delta x}$ by induction, it suffices to define $u_{\Delta x}(x,t)$ for $t_{j-1} < t \leq t_j$. For $t_{j-1} < t < t_j$, let $u_{\Delta x}(x,t)$ be obtained by solving the Riemann problems posed at time $t = t_{j-1}$ as in Theorem 3. Note that since all wave speeds are bounded by c, we assume that

$$\Delta x/\Delta t \geq 2c,$$

in order to insure that waves do not interact within one time step. Now to *re-pose* the constant states and the corresponding Riemann problems at time level t_j in the approximate solution, let $\mathbf{a} \equiv \{a_j\} \in \mathbf{A}$ denote a (fixed) random sequence, $0 < a_j < 1$, where $\mathbf{A}$ denotes the infinite product of intervals $(0,1)$ endowed with Lebesgue measure, $j = 0, 1, 2, \ldots$. Then define

$$u_{\Delta x}(x, t_j+) = u_{ij}, \quad for \ \ x_i \leq x < x_{i+1},$$

where

$$u_{ij} = u_{\Delta x}(x_i + a_j \Delta x, t_j-).$$

This completes the definition of the approximate solution $u_{\Delta x}$ by induction. (Note that $u_{\Delta x}$ depends on the choice of $\mathbf{a} \in \mathbf{A}$.) The restriction on the random sequence $\mathbf{a}$ will come at the end. The important point is that the waves in the solution $u_{\Delta x}$ at mesh point (x_i, t_j) solve the Riemann problem with $u_L = u_{\Delta x}(x_{i-1} + a_j \Delta x, t_j-)$ and right state $u_R = u_{\Delta x}(x_i + a_j \Delta x, t_j-)$, these being states that appear in the waves of the previous time level. In the case of system (2.1.15), we modify the above definition of wave strength $|\gamma|$ by defining

$$|\gamma| = |\ln(\rho_R) - \ln(\rho_L)|, \tag{2.7.94}$$

where u_L is the left state and u_R the right state of the wave γ. The proof of Theorem 1 is based on the following lemma:

Lemma 6 *Let u_L, u_M and u_R denote three arbitrary states satisfying (2.6.89) and (2.6.90). Let α_i, β_i and γ_i denote the waves that solve the Riemann problems $[u_L, u_M], [u_M, u_R]$ and $[u_L, u_R]$, respectively, $i = 1, 2$. Then*

$$|\gamma_1| + |\gamma_2| \leq |\alpha_1| + |\alpha_2| + |\beta_1| + |\beta_2|. \tag{2.7.95}$$

Proof: Lemma 6 follows from the special geometry of the shock curves that was obtained in Section 5. The important point is that the graphs of the shock curves $\mathbf{S}_i(u_L)$ in rs-plane (the plane of Riemann invariants (2.5.73) and (2.5.74)) have the same geometric shape, independent of u_L; and the 1-shock curves are reflections of the 2-shock curves about the line $r = s$. It is also important that Lemma 5 holds: namely, $0 \leq |ds/dr| < 1$ all along the 1-shock curves $\mathbf{S}_1$, and $0 < |dr/ds| \leq 1$ along $\mathbf{S}_2$. In particular, these latter conditions imply that the interaction of shock waves of the same family produces a stronger shock in the same family, together with a rarefaction wave of the opposite family. This special geometry is enough to imply that the sum of the strengths of the waves as projected onto the $(s - r)$-axis (recall that $s - r = 2K_0 \ln(\rho)$) is the same in the outgoing waves as in the incoming waves for any interaction $[u_L, u_M], [u_M, u_R] \to [u_L, u_R]$-except in the case when waves of the same family cancel, in which case the sum decreases. Note that Lemma 5 guarantees that the strength $|\gamma|$ is monotone in u_R along $S_i, i = 1, 2$, and this is needed to ensure that the Euclidean strength $|u_L - u_R|$ across a wave γ can be bounded by the strength of the wave as defined in (2.7.94). The best way to verify (2.7.95) in two wave interactions, the general case being a straightforward generalization of these, and we omit the details. [22, 19].□

Now let γ_{ij}^p denote the p-wave that appears in the solution of the Riemann problem posed at the mesh point (x_i, t_j) in the approximate solution $u_{\Delta x}, p = 1, 2$.

Lemma 7 *Let $u_{\Delta x}$ denote an approximate solution of the random choice method for initial data satisfying (2.6.89) and (2.6.90). Then*

$$\sum_{-\infty<i<\infty,p=1,2} |\gamma^p_{i,j+1}| \leq \sum_{-\infty<i<\infty,p=1,2} |\gamma^p_{ij}|. \qquad (2.7.96)$$

Proof: Consider the portions of the elementary waves $\gamma^p_{i-1,j}$, γ^p_{ij} and $\gamma^p_{i+1,j}$ that, at time level $t = t_{j+1}$, lie within the spacial interval $[x_{i-1}+a_{j+1}\Delta x, x_i + a_{j+1}\Delta x)$. Let these waves be denoted $\alpha^p_{i-1,j}$, $\bar{\gamma}^p_{i,j}$ and $\beta^p_{i+1,j}$ according to whether the wave emanates from the mesh point (x_{i-1},t_j), (x_i,t_j) or (x_{i+1},t_j), respectively. Then the waves $\alpha^p_{i-1,j}, p = 1,2$, are either all zero strength waves, (the random point lies closer to $x = x_i$ than the states in the waves $\alpha^p_{i-1,j}$ at time level $t = t_{j+1}$)), or else they are the waves that solve the Riemann problem with left state $u_L \equiv u_{\Delta x}(x_{i-1} + a_{j+1}\Delta x, t_{j+1}-)$, and right state $u_{i-1,j}$; the waves $\beta^p_{i+1,j}$ are similarly all zero or else they solve the Riemann problem with left state u_{ij} and right state $u_R \equiv u_{\Delta x}(x_i+a_{j+1}\Delta x, t_{j+1}-)$; and $\bar{\gamma}^p_{ij}$ solve the Riemann problem with left state either u_L or $u_{i-1,j}$ and right state either u_{ij} or u_L, depending on where the random points $(x_{i-1}+a_{j+1}\Delta x, t_{j+1}-)$ and $(x_i+a_{j+1}\Delta x, t_{j+1}-)$ fall relative to the waves γ^p_{ij} at time leve t_{j+1}. Note that in this notation, the waves $\gamma^p_{i,j+1}$ solve the Riemann problem $[u_L, u_R]$. Thus by Lemma 6 we can conclude that

$$\sum_{p=1,2} |\gamma^p_{i,j+1}| \leq \sum_{p=1,2} \{|\alpha^p_{i-1,j}| + |\bar{\gamma}^p_{ij}| + |\beta^p_{i+1,j}|\}. \qquad (2.7.97)$$

But summing i from $-\infty$ to $+\infty$ in (2.7.97) and rearranging terms gives

$$\sum_{p,i} |\gamma^p_{i,j+1}| \leq \sum_{p,i}\{|\alpha^p_{ij}| + |\bar{\gamma}^p_{ij}| + |\beta^p_{ij}|\} \leq \sum_{p,i} |\gamma^p_{ij}|,$$

the latter inequality holding because, by construction,

$$|\alpha^p_{ij}| + |\bar{\gamma}^p_{ij}| + |\beta^p_{ij}| = |\gamma^p_{ij}|.$$

This completes the proof of Lemma 6. $\square$

We now complete the proof of Theorem 1. First, since $\ln(\rho)$ is monotone along the wave curves $\mathbf{T_i}$, it follows that at time level $t \in (t_j, t_{j+1})$ in an approximate solution $u_{\Delta x}$, the sum of the strengths of the waves at time t is equivalent to the total variation in $\ln(\rho)$ of the approximate solution at time t:

$$Var\{\ln(\rho_{\Delta x}(\cdot, t))\} = \sum_{p,i} |\gamma^p_{ij}|. \qquad (2.7.98)$$

Thus, in the approximate solution,

$$Var\{\ln(\rho_{\Delta x}(\cdot, t+))\} \leq Var\{\ln(\rho_{\Delta x}(\cdot, s+))\}, \qquad (2.7.99)$$

whenever $s \leq t$. We now note that by Helly's theorem [26], L^1 limits of functions of uniformly bounded variation satisfy the same variation bound. This gives the first inequality (2.1.20) of Theorem 1 for any weak solution $u(x,t) = u(\rho(x,t), v(x,t))$ obtained as an L^1 limit of approximate solutions $u_{\Delta x}$ as $\Delta x \to 0$. Thus we show that (2.1.20) of Theorem 1 is a consequence of the following lemma :

Lemma 8 *(Glimm, [10]) Assume that the approximate solution $u_{\Delta x}$ satisfies*

$$Var\{u_{\Delta x}(\cdot, t)\} < V < \infty \tag{2.7.100}$$

for all $t \geq 0$. Then there exists a subsequence of mesh lengths $\Delta x \to 0$ such that $u_{\Delta x} \to u$ where $u(x, t)$ also satisfies (2.7.100). The approximate solutions converge pointwise a.e., and in L^1_{loc} at each time, uniformly on bounded x and t sets. Moreover, there exists a set $\mathbf{N} \subset \mathbf{A}$ of Lebesgue measure zero such that, if $\mathbf{a} \in \mathbf{A} - \mathbf{N}$, then $u(x, t)$ is a weak solution of the initial value problem (2.1.15), (2.1.16).

Using Lemma 8, the proof of (2.1.20) and (2.1.21) of Theorem 1 is completed once we show that the estimates (2.1.18) and (2.1.19) imply (2.7.100) for the approximate Glimm scheme solutions. For this, note that (2.1.18) and (2.1.19) imply that there exist states $\rho_\infty = \lim_{x\to\infty} \rho_0(x)$ and $u_\infty = \lim_{x\to\infty} u_0(x)$. But (2.1.19) implies that the total variation in $\ln\left\{\frac{c+v}{c-v}\right\}$ is finite at time $t = 0+$ in the approximate solution $u_{\Delta x}$, and thus

$$Var\{\ln(\rho_{\Delta x}(\cdot, 0+))\} < V_0,$$

where V_0 depends only on the initial total variation bounds in (2.1.18) and (2.1.19). Thus, by (2.7.99),

$$Var\{\ln(\rho_{\Delta x}(\cdot, t+))\} < V_0, \tag{2.7.101}$$

for all positive times $t > 0$. But it follows directly from (2.5.86) and (2.5.85) of Lemma 5 that the variation in $\ln\{\frac{c+v}{c-v}\}$ is bounded uniformly by the variation in $\ln(\rho)$ across every elementary wave in each approximate solution $u_{\Delta x}$. Thus (2.7.101) implies that

$$Var\left\{\ln\left(\frac{c+v}{c-v}\right)(\cdot, t+)\right\} < V_1, \tag{2.7.102}$$

at each $t > 0$ in an approximate solution $u_{\Delta x}$, where V_1 depends only on V_0. It now follows from (2.7.101) and (2.7.102) that for every Δx and $\mathbf{a} \in \mathbf{A}, \rho_\infty = \lim_{x\to\infty} \rho_{\Delta x}(x, t)$ and $v_\infty = \lim_{x\to\infty} v_{\Delta x}(x, t)$. This means that (ρ_∞, v_∞) is a constant state appearing in the approximate solutions at each fixed time level. This together with (2.7.101) and (2.7.102) implies that there exists a constant $M > 0$ such that

$$1/M < \rho_{\Delta x}(x, t) < M,$$

and

$$-c + 1/M < v_{\Delta x}(x,t) < c - 1/M,$$

for all x and $t \geq 0$, uniformly in Δx. The desired result (2.7.100) follows from these latter two bounds because, in light of Proposition 1, the Jacobian determinant $\left|\frac{\partial(u^0, u^1)}{\partial(\rho, v)}\right|$ is bounded away from zero (uniformly in Δx) on the image of the approximate solutions $u_{\Delta x}$. Thus by Lemma 8 there exists a subsequence of mesh lengths $\Delta x \to 0$ such that $u_{\Delta x} \to u(\rho(x,t), v(x,t))$ where u satisfies (2.7.100), (2.1.20) and (2.1.21), and the convergence is pointwise a.e., and in L^1_{loc} at each time, uniformly on bounded x and t sets. By Proposition 2, V_0 and V_1 must be Lorentz invariant constants.$\square$

2.8 Appendix

In this section we show that the transformation properties of characteristics and shocks in a relativistic system of conservation laws follows directly from the covariance properties of the Rankine-Hugoniot jump relations. In particular, we show that the characteristic curves and shock curves associated with a system of conservation laws $divT = 0$, transform, under under general nonlinear spacetime coordinate transformations, like the level curve of a scalar function. We assume that the divergence is taken with respect to a Lorentz metric g defined in four dimensional spacetime. As a consequence, we show that the wave speeds λ_i and the shock speeds s_i, defined for systems (2.1.10) and (2.1.15), transform according to the special relativistic velocity transformation law (2.2.28).

Thus, if g_{ij} denotes a fixed Lorentzian metric defined on a four dimensional spacetime $\mathbf{x} \equiv (x^0, ..., x^3), x^0 = ct$, let Γ^i_{jk} denote the Christoffel symbols which define the unique symmetric connection associated with g_{ij}, namely

$$\Gamma^i_{jk} \equiv \frac{1}{2} g^{i\sigma} \{g_{\sigma j,k} + g_{k\sigma,j} - g_{jk,\sigma}\}.$$

Let T_{ij} be a symmetric $(0, 2)$-tensor which we take to be the stress energy tensor for some field in spacetime. Conservation of energy-momentum based on the metric g then reads $divT = 0$, where the covariant divergence is given in coordinates by [35, 8]

$$divT \equiv T^\sigma_{j;\sigma} \equiv T^\sigma_{j,\sigma} + \Gamma^\sigma_{\tau\sigma} T^\tau_j - \Gamma^\tau_{j\sigma} T^\sigma_\tau. \tag{2.8.103}$$

Here, ",i" denotes $\partial/\partial x^i$ and ";i" denotes the covariant derivative.

Definition 1 *Assume that T_{ij} (defined in a given coordinate system $\mathbf{x}$) is smooth except for a jump discontinuity across a smooth surface $\phi(\mathbf{x}) = 0$, where $d\phi \equiv n_\sigma dx^\sigma \neq 0$. Then we say that T is a <u>weak solution</u> of $divT = 0$*

if this equation holds away from the surface $\phi(\mathbf{x}) = 0$, *and across the surface the following (Rankine-Hugoniot) jump conditions hold:*

$$[T_j^\sigma n_\sigma] = 0. \tag{2.8.104}$$

Here, as usual, the square brackets around a quantity denote the jump in the quantity across the surface $\phi = 0$,

$$[T_j^\sigma n_\sigma] \equiv (T_j^\sigma n_\sigma)_R - (T_j^\sigma n_\sigma)_L = [T_j^\sigma] n_\sigma.$$

One can show that the jump relation 2.8.104 is implied by the weak formulation of $divT = 0$ in the sense of the theory of distributions [26], and implies conservation of the physical energy-momentum across the surface of discontinuity $\phi = 0$. The equivalence of the weak formulation follows from integration by parts, observing that the non-divergence terms $\Gamma_{\tau\sigma}^\sigma T_j^\tau - \Gamma_{j\sigma}^\tau T_\tau^\sigma$ contain no derivative of T.

From the jump conditions we obtain the following proposition:

Proposition A 1 *A shock surface transforms (under arbitrary nonlinear changes of spacetime coordinates) as the level curve of a scalar function defined on spacetime.*

Proof: If $[T_j^i] n_i = 0$ holds on the surface $\phi = 0$ in one coordinate system x where $d\phi = n^i dx^i$ then it holds in every other coordinate system because T transforms like a (1,1)-tensor, and $d\phi$ is a 1-form.□

Now restricting to a 2−dimensional Minkowski spacetime, so that $g_{ij} \equiv \eta_{ij}$, Proposition A1 implies that the shock curve has tangent vector $d\mathbf{x}/d\tau = X$ if and only if $n_i dx^i(X) = 0$, where $n_i dx^i = d\phi$ and ϕ is a function constant along the shock curve. Thus, letting X^i denote the components of X, we conclude that the shock speed $s = dx^1/dt$ is given by

$$s = c\frac{X^1}{X^0} = -c\frac{n_0}{n_1}.$$

Proposition A 2 *Under Lorentz transformations, the shock speeds s transform according to the velocity transformation rule (2.2.28).*

Proof: Consider a Lorentz transformation taking the unbarred coordinates x^i to the barred coordinates $\bar{x}^\alpha$, such that the barred frame moves with velocity μ as measured in the unbarred frame x^i. Then

$$\bar{x}^\alpha = \Lambda(\mu)_i^\alpha x^i,$$

where

$$\Lambda(\mu)_i^\alpha = \begin{bmatrix} cosh(\theta) & sinh(\theta) \\ sinh(\theta) & cosh(\theta) \end{bmatrix}, \tag{2.8.105}$$

and $tanh(\theta) = \mu/c$, [35]. Then the vector X, tangent to the shock curve, transforms as

$$\bar{X}^\alpha = \Lambda(\mu)^\alpha_i X^i.$$

Thus in the barred coordinates, the shock speed is given by

$$\bar{s} = c\frac{\bar{X}^1}{\bar{X}^0} = c\frac{X^0 cosh(\theta) + X^1 sinh(\theta)}{X^0 sinh(\theta) + X^1 cosh(\theta)} = \frac{\mu + s}{1 + \frac{\mu v}{c^2}}.$$

This completes the proof of the Proposition A2.□

Proposition A 3 *The characteristic curves for a system of conservation laws (2.2.24) transform as level curves of functions in physical space. Moreover, under Lorentz transformations, the wave speeds λ_i (the eigenvalues of dF) transform according to the velocity transformation law (2.2.28).*

Proof: By Theorem 2, $\lambda_i = \lim_{\varepsilon \to 0} s_i(\varepsilon)$. Thus, by continuity, λ_i transforms as a velocity (2.2.28) under Lorentz transformations because $s_i(\varepsilon)$ does for each fixed ε. Since the characteristic curves are given by $dx^1/dt = \lambda_i$ and λ_i transforms as a velocity, it follows that the characteristic curves must transform like level curves of functions. Thus by duality, in two dimensional spacetime, the tangents to the shock curves and characteristic curves transform like vectors.□

3

A Shock Wave Formulation of the Einstein Equations

3.1 Introduction

In this chapter we show that Einstein equations (1.3.2)-(1.3.5) are weakly equivalent to the system of conservation laws with time dependent sources (1.4.3),(1.4.4), so long as the metric is in the smoothness class $C^{0,1}$, and T is in L^∞. Inspection of equations (1.3.2)-(1.3.5) shows that it is in general *not possible* to have metrics smoother than Lipschitz continuous, (that is, smoother than $C^{0,1}$ at shocks), when the metric is written in standard Schwarzschild coordinates. Indeed, at a shock wave, the fluid variables, and hence T, suffer jump discontinuities. At such a discontinuity, (1.3.2)-(1.3.5) imply that A_r, B_r and B_t all suffer jump discontinuities as well.

We restrict to spacetime metrics g that are spherically symmetric, by which we mean that g lies within the general class of metrics that take the form,

$$ds^2 = g_{ij}dx^i dx^j \equiv -A(r,t)dt^2 + B(r,t)dr^2 + 2D(r,t)dtdr + C(r,t)d\Omega^2,$$
(3.1.1)

where the components A, B, C and D of the metric are assumed to be functions of the radial and time coordinates r and t alone, $d\Omega^2 \equiv d\theta^2 + \sin^2(\theta)d\phi^2$ denotes the line element on the 2-sphere, and $x \equiv (x^0, ..., x^3) \equiv (t, r, \theta, \phi)$, denotes the underlying coordinate system on spacetime, [35, 34, 14, 21]. In this case we assume that the 4-velocity $\mathbf{w}$ is radial, by which we mean that the x-components of $\mathbf{w}$ are given by

$$w^i = (w^0(r,t), w^1(r,t), 0, 0), \quad i = 0, ..., 3, \quad \text{respectively,}$$
(3.1.2)

for some functions w^0 and w^1.

Now it is well known that in general there exists a coordinate transformation $(r, t) \rightarrow (\bar{r}, \bar{t})$ that takes an arbitrary metric of form (3.1.1) over to one of form, [35],

$$ds^2 = g_{ij}dx^i dx^j \equiv -A(r,t)dt^2 + B(r,t)dr^2 + r^2 d\Omega^2. \tag{3.1.3}$$

A metric of form (3.1.3) is said to be in the standard Schwarzschild coordinates, or (standard coordinate guage), and it is our purpose here to establish the weak formulation of the Einstein equations (1.3.2)-(1.3.5) for metrics of the form (3.1.3) in the case when A and B are finite, and satisfy $AB \neq 0$.

The general problem of making sense of gravitational metrics that are only Lipschitz continuous at shock surfaces was taken up in [29]. The analysis there identifies conditions that must be placed on the metric in order to insure that conservation holds at the shock, and that there do not exist delta-function sources at the shock, [15]. When these conditions are met, the methods in [29] imply the existence of a $C^{1,1}$ coordinate transformation that improves the level of smoothness of the metric components from $C^{0,1}$ up to $C^{1,1}$ at the shock. However, the results in [29] apply only to smooth interfaces that define a single shock surface for which $G = \kappa T$ holds identically on either side. For general shock wave solutions of the form (3.1.3), (that can contain multiplicities of interacting shock waves), it is an open question whether there exists a coordinate transformation, (say to a metric in the more general class (3.1.1)), that can increase the level of smoothness of the metric components by one order. For this reason, we now show that the mapping $(r,t) \rightarrow (\bar{r},\bar{t})$ that takes an arbitrary metric of form (3.1.1) over to one of form (3.1.3), implies a loss of one order of differentiability in the metric components when shock waves are present. This argues that our results are *consistent* with the existence of such a smoothing coordinate transformation, but still leaves open the problem of the *existence* of such a transformation.

Thus we now review the construction of the mapping $(r,t) \rightarrow (\bar{r},\bar{t})$ that takes an arbitrary metric of form (3.1.1) over to one of form (3.1.3), [35]. To start, one must assume that the metric component $C(t,r)$ in (3.1.1) satisfies the condition that for each fixed t, C increases from zero to infinity as r increases from zero to infinity, and that

$$\frac{\partial}{\partial r}C(r,t) \neq 0. \tag{3.1.4}$$

(These are not unreasonable assumptions considering that C measures the areas of the spheres of symmetry.) Define

$$\bar{r} = \sqrt{C(r,t)}. \tag{3.1.5}$$

Then the determinant of the Jacobian of the mapping $(r,t) \rightarrow (\bar{r},t)$ satisfies

$$\left|\frac{\partial \bar{r}}{\partial r}\right| = \frac{\partial}{\partial r}\sqrt{C(r,t)} \neq 0,$$

in light of (3.1.4). Thus the transformation to $(\bar{r},t)$ coordinates is (locally) a nonsingular transformation, and in $(\bar{r},t)$ coordinates the metric (3.1.1) takes the form

$$ds^2 = -A(r,t)dt^2 + B(r,t)dr^2 + 2E(r,t)dtd\bar{r} + r^2 d\Omega^2. \qquad (3.1.6)$$

(Here we have replaced $\bar{r}$ by r and A, B and E stand in for the transformed components.) It is easy to verify that, to eliminate the mixed term, it suffices to define the time coordinate $\bar{t}$ so that, cf. [35],

$$d\bar{t} = \phi(r,t)\left\{A(r,t)dt - E(r,t)dr\right\}. \qquad (3.1.7)$$

In order for (3.1.7) to be exact, so that $\bar{t}$ really does define a coordinate function, the integrating factor ϕ must be chosen to satisfy the (linear) PDE

$$\frac{\partial}{\partial r}\left\{\phi(r,t)A(r,t)\right\} = -\frac{\partial}{\partial t}\left\{\phi(r,t)E(r,t)\right\}. \qquad (3.1.8)$$

But we can solve (3.1.8) for $\phi(r,t)$ from initial data $\phi(r,t_0)$, by the method of characteristics. From this it follows that, (at least locally), we can transform metrics of form (3.1.1) over to metrics of form (3.1.3) by coordinate transformation. To globalize this procedure, we need only assume that $C_r(t,r) \neq 0$, and that C takes values from zero to infinity at each fixed t. Now note that in general $\phi(r,t)$, the solution to (3.1.8), will have the same level of differentiability as $A(r,t)$ and $E(r,t)$; and so it follows that the components of dt and dr in (3.1.7) will have this same level of differentiability. This implies that the $\bar{t}$ transformation defined by (3.1.7) preserves the level of smoothness of the metric component functions. On the other hand, the $\bar{r}$ transformation in (3.1.5) reduces the level of differentiablility of the metric components by one order. Indeed, the level of smoothness of the transformed metric component functions are in general no smoother than the Jacobian that transforms them, and by (3.1.5), the Jacobian of the transformation contains the terms C_r and C_t which will in general be only $C^{0,1}$ when $C \in C^{1,1}$. Thus, if we presume, (motivated by [28]), that for general spherically symmetric shock wave solutions of $G = \kappa T$, that there exists a coordinate system in which the metric takes the form (3.1.1), and the components of g in these coordinates are $C^{1,1}$ functions of these coordinates, then it follows that we cannot expect the transformed metrics of form (3.1.3) to be better than $C^{0,1}$, that is, Lipschitz continuous. The equations we derive below allow for metrics in the smoothness class $C^{0,1}$, but in general they do not admit solutions smoother than Lipschitz continuous. It remains an open question whether solutions to these equations can be smoothed by coordinate transformation when shock waves are present.

In Section 3 we verify the equivalence of several weak formulations of the Einstein equations that allow for shock waves, and that are valid for metrics of form (3.1.3), in the smoothness class $C^{0,1}$. In Section 4, we show that these equations are weakly equivalent to the system (1.4.3)-(1.4.4) of conservation laws with time dependent sources. In the next chapter, we give an existence theory for these equations with general Cauchy data of bounded variation, thereby demonstrating the consistency of the Einstein equations for weak (shock wave) solutions within the class of $C^{0,1}$ metrics.

3.2 The Einstein Equations for a Perfect Fluid with Spherical Symmetry

In this section we study the system of equations obtained from the Einstein equations under the assumption that the spacetime metric g is spherically symmetric. So assume that the gravitational metric g is of the form (3.1.1), and to start, assume that T^{ij} is any arbitrary stress tensor. To obtain the equations for the metric components A and B implied by the Einstein equations (1.2.1), plug the ansatz (3.1.3) into the Einstein equations (1.2.1). The resulting system of equations (1.3.2)-(1.3.5) is obtained using MAPLE. Equations (1.3.2)-(1.3.5) represent the (0,0), (0,1), (1,1) and (2,2) components of $G^{ij} = \kappa T^{ij}$, respectively, (as indexed by T on the RHS of each equation). The (3,3) equation is a multiple of the (2,2) equation, and all remaining components are identically zero. (Note that MAPLE defines the curvature tensor to be minus one times the curvature tensor defined in (1.1.14).)

We are interested in solutions of (1.3.2)-(1.3.5) in the case when shock waves are present. Since A and B have discontinuous derivatives when shock waves are present, it follows that (1.3.5), being second order, cannot hold classically, and thus equation (1.3.5) must be taken in the weak sense, that is, in the sense of the theory of distributions. To get the weak formulation of (1.3.5), multiply through by AB^2 to clear away the coefficient of the highest (second) order derivatives, then multiply through by a test function and integrate the highest order derivatives once by parts. It follows that if the test function is in the class $C_0^{1,1}$, (that is, one continuous derivative that is Lipschitz continuous, the subscript zero denoting compact support), and if the metric components A and B are in the class $C^{0,1}$, and T^{ij} is in class L^∞, then all terms in the integrand of the resulting integrated expression are at most discontinuous, and so all derivatives make sense in the classical pointwise a.e. sense.

In order to account for initial and boundary conditions in the weak formulation, it is standard to take the test function ϕ to be nonzero at $t = 0$ or at the specified boundary. In this case, when we integrate by parts to obtain the weak formulation, the boundary integrals are non-vanishing, and their inclusion in the weak formulation represents the condition that the boundary values are taken on in the weak sense. Thus, for example, if the boundary is $r = r_0 \geq 0$, we say $\phi \in C_0^{1,1}(r \geq r_0, t \geq 0)$ to indicate that ϕ can be nonzero initially and at the boundary $r = r_0$, thereby implicitly indicating that boundary integrals will appear in the weak formulation based on such test functions.

We presently consider various equivalent weak formulations of equations (1.3.2)-(1.3.5), and we wish to include the equivalence of the weak formulation of boundary conditions in the discussion. Thus, in order to keep things as simple as possible, *we now restrict to the case* of weak solutions of (1.3.2)-(1.3.5) defined on the domain $r \geq r_0 \geq 0$, $t \geq 0$, and we always assume that test functions ϕ lie in the space $\phi \in C_0^{1,1}(t \geq 0, r \geq r_0)$ so that initial and boundary values are accounted for in the weak formulation. (This is the simplest case in which to rigorously demonstrate the equivalence of several weak formulations of initial boundary value problems. More general domains can be handled in a similar manner.)

Note that because (1.3.2)-(1.3.4) involve only first derivatives of A and B, and $A, B \in C^{0,1}$, it follows that (1.3.2)-(1.3.4) can be taken in the strong sense, that is, derivatives can be taken in the pointwise a.e. sense. The continuity of A and B imply also that the initial and boundary values are taken on strongly in any $C^{0,1}$ weak solution of $(1.3.2) - (1.3.4)$. On the other hand, equation (1.3.5) involves second derivatives, and so this last equation is the only one that requires a weak formulation. The weak formulation of (1.3.5) is thus obtained on domain $t \geq 0$, $r \geq r_0 \geq 0$ by multiplying through by a test function $\phi \in C_0^{1,1}(r \geq r_0, t \geq 0)$ and integrating by parts. This yields the following weak formulation of (1.3.5):

$$
0 = \int_{r_0}^{\infty} \int_0^{\infty} \left\{ -\frac{B_t \phi_t}{rAB^2} - \frac{B_t \phi}{r} \left(-\frac{A_t}{A^2 B^2} - \frac{2B_t}{AB^3} \right) + \frac{A' \phi'}{rAB^2} \right.
$$
$$
+ A' \phi \left(-\frac{1}{r^2 AB^2} - \frac{A'}{rA^2 B^2} - \frac{2B'}{rAB^3} \right) + \frac{\phi}{rAB^2}\Phi + \frac{2\kappa r}{B}\phi T^{22} \Bigg\} \, dr dt
$$
$$
- \int_{r_0}^{\infty} \frac{B_t(r,0)\phi(r,0)}{rA(r,0)B^2(r,0)} \, dr + \int_0^{\infty} \frac{A'(r_0,t)\phi(r_0,t)}{r_0 A(r_0,t)B^2(r_0,t)} \, dt \qquad (3.2.1)
$$

Our first proposition states that the weak formulation (3.2.1) of equation (1.3.5) may be replaced by the weak formulation of the conservation laws $divT = 0$, so long as A and B are in $C^{0,1}$ and $T^{ij} \in L^{\infty}$.

Proposition 3 *Assume that $A, B \in C^{0,1}(r \geq r_0, t \geq 0)$, $T^{ij} \in L^{\infty}(r \geq r_0, t \geq 0)$ and assume that A, B and T solve (1.3.2)-(1.3.4) strongly. Then A, B and T solve $T^{1i}_{;i} = 0$, (the 1-component of $DivT = 0$), weakly if and only if A, B and T satisfy (3.2.1).*

Proof: The proof strategy is to modify (3.2.1) and the weak form of conservation using (1.3.2)-(1.3.4) as identities, and then observe that the two are identical at an intermediate stage. To begin, substitute for B_t and A' in several places in (3.2.1) to obtain the equivalent condition

$$
0 = \int_{r_0}^{\infty} \int_0^{\infty} \left\{ \kappa T^{01} \varphi_t + \kappa T^{11} \varphi' + \frac{\partial}{\partial r}\left(\varphi \frac{(B-1)}{r^2 B^2} \right) + \varphi \left[-\frac{\partial}{\partial r}\left(\frac{B-1}{r^2 B^2} \right) \right. \right.
$$
$$
+ \frac{B_t}{r}\left(\frac{A_t}{A^2 B^2} + \frac{2B_t}{AB^3} \right) + A'\left(-\frac{1}{r^2 AB^2} - \frac{A'}{rA^2 B^2} - \frac{2B'}{rAB^3} \right)
$$
$$
+ \frac{1}{rAB^2}\Phi + \frac{2\kappa r}{B}T^{22} \Bigg] \Bigg\} \, dr \, dt
$$
$$
+ \kappa \int_{r_0}^{\infty} T^{01}(r,0)\varphi(r,0) \, dr + \kappa \int_0^{\infty} \varphi(r_0,t) \left[T^{11}(r_0,t) \frac{B(r_0,t)-1}{r_0^2 B^2(r_0,t)} \right] dt
$$

$$= \int_{r_0}^{\infty} \int_0^{\infty} \left\{ \kappa T^{01} \varphi_t + \kappa T^{11} \varphi' + \varphi \left[\frac{B'(B-2)}{r^2 B^3} + 2 \frac{(B-1)}{r^3 B^2} \right. \right.$$

$$+ \frac{B_t}{r} \left(\frac{A_t}{A^2 B^2} + \frac{2B_t}{AB^3} \right) + A' \left(-\frac{1}{r^2 A B^2} - \frac{A'}{rA^2 B^2} - \frac{2B'}{rAB^3} \right)$$

$$\left. \left. + \frac{1}{rAB^2} \Phi + \frac{2\kappa r}{B} T^{22} \right] \right\} dr \, dt$$

$$+ \kappa \int_{r_0}^{\infty} T^{01}(r,0) \varphi(x,0) \, dr + \kappa \int_0^{\infty} \varphi(r_0,t) T^{11}(r_0,t) \, dt. \tag{3.2.2}$$

Now, the weak form of conservation of energy-momentum is given by

$$0 = \int_{r_0}^{\infty} \int_0^{\infty} \left\{ T^{01} \varphi_t + T^{11} \varphi' - \left(\Gamma_{i0}^i T^{01} + \Gamma_{i1}^i T^{11} \right. \right. \tag{3.2.3}$$

$$\left. + \Gamma_{00}^1 T^{00} + 2\Gamma_{01}^1 T^{01} + \Gamma_{11}^1 T^{11} + 2\Gamma_{22}^1 T^{22} \right) \varphi \right\} dr \, dt$$

$$+ \int_{r_0}^{\infty} T^{01}(r,0) \varphi(x,0) \, dr + \int_0^{\infty} \varphi(r_0,t) T^{11}(r_0,t) \, dt.$$

Here, we have used the fact that $T^{22} = \sin^2 \theta T^{33}$, $T^{ij} = 0$ if $i \neq j = 2$ or 3, and $\Gamma_{33}^1 = \sin^2 \theta \Gamma_{22}^1$. Next, we calculate the connection coefficients Γ_{jk}^i using (1.1.10) to obtain,

$$\begin{array}{ll}
\Gamma_{i0}^i = \frac{1}{2} \left(\frac{A_t}{A} + \frac{B_t}{B} \right) & \Gamma_{i1}^i = \frac{1}{2} \left(\frac{A'}{A} + \frac{B'}{B} + \frac{4}{r} \right) \\
\Gamma_{00}^0 = \frac{A_t}{2A} & \Gamma_{01}^0 = \frac{A'}{2A} \\
\Gamma_{11}^0 = \frac{B_t}{2A} & \Gamma_{22}^0 = 0 = \Gamma_{33}^0 \\
\Gamma_{00}^1 = \frac{A'}{2B} & \Gamma_{01}^1 = \frac{B_t}{2B} \\
\Gamma_{11}^1 = \frac{B'}{2B} & \Gamma_{22}^1 = -\frac{r}{B} \\
\Gamma_{33}^1 = -\frac{r \sin^2 \theta}{B}.
\end{array} \tag{3.2.4}$$

Substituting the above formulas for Γ_{jk}^i into (3.2.3) and using (1.3.2)-(1.3.4) as identities to eliminate some of the T^{ij} in favor of expressions involving A, B and r, we see that (3.2.3) is equivalent to:

$$0 = \int_{r_0}^{\infty} \int_0^{\infty} \left\{ T^{01} \varphi_t + T^{11} \varphi' + \frac{\varphi}{\kappa} \left[\frac{1}{2} \left(\frac{A_t}{A} + \frac{3B_t}{B} \right) \frac{B_t}{rAB^2} \right. \right.$$

$$- \frac{1}{2} \left(\frac{A'}{A} + \frac{2B'}{B} + \frac{4}{r} \right) \frac{1}{r^2 B^2} \left(r\frac{A'}{A} - (B-1) \right) \tag{3.2.5}$$

$$\left. \left. - \frac{A'}{2r^2 AB} \left(r\frac{B'}{B} + (B-1) \right) + 2\kappa \frac{r}{B} T^{22} \right] \right\} dr \, dt$$

$$+ \int_{r_0}^{\infty} T^{01}(r,0) \varphi(r,0) \, dr + \int_0^{\infty} \varphi(r_0,t) T^{11}(r_0,t) \, dt.$$

After some simplification, it is clear that (3.2.2) is equal to (3.2.5). This completes the proof of Proposition 3. $\square$

We next show that the Einstein equations (1.3.2)-(1.3.4) together with $DivT = 0$ are overdetermined. Indeed, we show that for weak solutions with Lipschitz continuous metric, either (1.3.2) or (1.3.3) may be dropped in the sense that the dropped equation will reduce to an identity on any solution of the remaining equations, so long as the dropped equation is satisfied by either the initial or boundary data, as appropriate. The following proposition addresses the first case, namely, for weak solutions in which the metric is Lipschitz continuous, the first Einstein equation (1.3.2) reduces to an identity on solutions of (1.3.3)-(1.3.4), so long as (1.3.2) is satisfied by the intial data.

Theorem 6 *Assume that $A, B \in C^{0,1}$ and $T \in L^\infty$ solve (1.3.3), (1.3.4) strongly, and solve $DivT = 0$ weakly. Then if $A, B,$ and T satisfy (1.3.2) at $t = 0$, then A, B, and T also solve (1.3.2) for all $t > 0$.*

Proof: We first give the proof for the case when A, B and T are assumed to be classical smooth solutions of (1.3.3), (1.3.4) and $DivT = 0$. This is followed by several lemmas necessary for the extension of this to the weak formulation, which is given in the final proposition. So to start, assume that $A, B,$ and T are all smooth functions, and thus solve $DivT = 0$ strongly. For the proof in this case, define

$$H^{ij} \equiv G^{ij} - \kappa T^{ij}. \tag{3.2.6}$$

Because (1.3.3) and (1.3.4) hold, $H^{01} \equiv H^{11} \equiv 0$. Since by assumption $T^{ij}_{;i} = 0$ and since $G^{ij}_{;i} = 0$ for any metric tensor as a consequence of the Bianchi identities, it follows that

$$0 = H^{ij}_{;i} = H^{ij}_{,i} + \Gamma^i_{ik} H^{kj} + \Gamma^j_{ik} H^{ik}. \tag{3.2.7}$$

In particular, setting $j = 0$,

$$0 = H^{i0}_{;i} = H^{i0}_{,i} + \Gamma^i_{ik} H^{k0} + \Gamma^0_{ik} H^{ik}. \tag{3.2.8}$$

By hypothesis, $H^{i0} = 0$ when $i \neq 0$. In addition, the connection coefficients Γ^0_{ik} are zero unless i or k equal 0 or 1. Therefore, (3.2.8) reduces to the linear ODE

$$0 = H^{00}_{,0} + \left(\Gamma^i_{i0} + \Gamma^0_{00} \right) H^{00}, \tag{3.2.9}$$

at each fixed r. By hypothesis, H^{00} is initially zero, and since we assume that H^{00} is a smooth solution of (3.2.9), it follows that H^{00} must continue to be zero for all $t > 0$.

Next, assume only that $A, B \in C^{0,1}$ and $T \in L^\infty$ so that (1.3.3), (1.3.4) hold strongly, (that is, in a pointwise a.e. sense), but that $DivT = 0$ is only known to hold weakly. In this case, the argument above has a problem because when $g \in C^{0,1}$, the Einstein tensor G, viewed as a second order operator on the metric components A and B, can only be defined weakly when A and B are only Lipschitz continuous. It follows that the Bianchi identities, and hence the identity $DivG = 0$, (which involves first order derivatives of the components of the curvature tensor), need no longer be valid even in a weak sense. Indeed, G can have delta function sources at an interface at which the metric is only Lipschitz continuous, c.f. [28]. However, the above argument only involves the 0'th component of $DivG = 0$, and the 0'th component of $DivG = 0$ involves only derivatives of the components G^{i0}, and as observed in (1.3.2), (1.3.3), these components only involve *first* derivatives of A and B. Specifically, the weak formulation of $G^{0i}_{;i} = 0$ is given by,

$$0 = \int_{r_0}^{\infty} \int_0^{\infty} \left\{ -\phi_i G^{i0} + \phi \left(\Gamma^i_{ik} G^{k0} + \Gamma^0_{ik} G^{ik} \right) \right\} dr dt \qquad (3.2.10)$$
$$- \int_{r_0}^{\infty} \phi(r,0) G^{00}(r,0) dr - \int_0^{\infty} \phi(r_0,t) G^{10}(r_0,t) dt,$$

and since, by (1.3.2), (1.3.3), G^{i0} involves only first order derivatives of A and B, it follows that the integrand in (3.2.10) is a classical function defined pointwise a.e. when $A, B \in C^{0,1}$. But (3.2.10) is *identically* zero for all *smooth* A and B because $DivG = 0$ is an identity. Thus, when $A, B \in C^{0,1}$, we can take a sequence of smooth functions $A_\varepsilon, B_\varepsilon$ that converge to A and B in the limit $\varepsilon \to 0$, (c.f. Theorem 7 below), such that the derivatives converge a.e. to the derivatives of A and B. It follows that we can take the limit $\varepsilon \to 0$ (3.2.10) and conclude that (3.2.10) continues to hold under this limit. Putting this together with the fact that $DivT = 0$ is assumed to hold weakly, we conclude that

$$H^{0i}_{;i} = (G^{0i} - T^{0i})_{;i} = 0,$$

in the weak sense, which means that H^{00} is in L^∞ and satisfies the condition

$$0 = \int_{r_0}^{\infty} \int_0^{\infty} \left\{ -\phi_0 H^{00} + \phi \left(\Gamma^i_{i0} + \Gamma^0_{00} \right) H^{00} \right\} dr dt \qquad (3.2.11)$$
$$- \int_{r_0}^{\infty} \phi(r,0) G^{00}(r,0) dr - \int_{r_0}^{\infty} \phi(r,0) H^{00}(r_0,t) dr.$$

Therefore, to complete the proof of Theorem 6, we need only show that if A, B and T solve (1.3.3), (1.3.4) classically and $DivT = 0$ weakly, then a weak L^∞ solution H^{00}, (i.e., that satisfies (3.2.11)), of (3.2.9) must be zero almost everywhere if it is zero initially. Thus it suffices to prove the following proposition:

Proposition 4 *Assume that $H, f \in L^{\infty}_{loc}(\mathbf{R} \times \mathbf{R})$. Then every L^{∞}_{loc} weak solution to the initial value problem*

$$H_t + fH = 0$$
$$H(x,0) = H_0(x). \tag{3.2.12}$$

with initial data $H_0 \equiv 0$ is unique, and identically equal to zero a.e., for all $t > 0$.

Proof: We use the following standard theorem, [9],

Theorem 7 *Let U be any open subset of $\mathbf{R}^n$. Then $u \in W^{1,\infty}_{loc}(U)$ if and only if u is locally Lipschitz continuous in U, in which case the weak derivative of u agrees with the classical pointwise a.e derivative as a function in $L^{\infty}_{loc}(U)$.*

Corollary 1 *Let u and f be real valued functions, $u, f : \mathbf{R} \to \mathbf{R}$, such that $u, f \in L^{\infty}[0,T]$, and u is a weak solution of the initial value problem*

$$u_t + fu = 0,$$
$$u(0) = 0, \tag{3.2.13}$$

on the interval $[0,T]$. Then $u(t) = 0$ for all $t \in [0,T]$.

Proof of Corollary: Statement (3.2.13) says that the distributional derivative u_t agrees with the L^{∞} function fu on the interval $[0,T]$, and thus we know that $u \in W^{1,\infty}_{loc}(0,t)$. Therefore, by Theorem 7, u is locally Lipschitz continuous on $(0,T)$, and the weak derivative u_t agrees with the pointwise a.e. derivative of u on $(0,T)$. Thus it follows from (3.2.13) that on any subinterval $[a,b]$ of $[0,T]$ on which $u \neq 0$, we must have

$$\frac{d}{dt}[\ln u] = \frac{u_t}{u} = -f, \quad a.e. \tag{3.2.14}$$

Moreover, since u is Lipschitz continuous, both u and $\ln(u)$ are absolutely continuous on $[a,b]$, so we can integrate (3.2.14) to see that

$$u(t) = u(a)e^{-\int_0^t f(\xi)d\xi}, \tag{3.2.15}$$

for all $t \in [a,b]$. But u is continuous, so (3.2.15) applies in the limit that a decreases to the first value of $t = t_0$ at which $u(t_0) = 0$. Thus (3.2.15) implies that $u(t) = 0$ throughout $[a,b]$, and hence we must have $u(t) = 0$ for all $t \in [0,T]$, and the Corollary is true.

The proof of Proposition 4 now follows because it is easy to show that if H is an L^{∞} weak solution of (3.2.12), then $H(x, \cdot)$ is a weak solution of the scalar ODE $H_t + fH = 0$ for almost every x. (Just factor the test functions into products of the form $\phi_1(t)\phi_2(x)$.)

Using Proposition 4, we see that if equation (1.3.2) holds on the initial data for a solution of (1.3.3), (1.3.4), and $DivT = 0$, then equation (1.3.2) will hold for all t. By a similar argument, it follows that if (1.3.3) holds for the boundary data of a solution to (1.3.2), (1.3.4), and $DivT = 0$, then (1.3.3) will hold for all r and t. We record this in the following theorem:

Theorem 8 *Assume that $A, B \in C^{0,1}$ and $T \in L^\infty$ solve (1.3.2), (1.3.4) strongly, and solve $DivT = 0$ weakly, in $r \geq r_0$, $t \geq 0$. Then if A, B, and T satisfy (1.3.3) at $r = r_0$, then A, B, and T also solve (1.3.2) for all $r > r_0$.*

3.3 The Einstein Equations as a System of Conservation Laws with Sources

Conservation of energy and momentum is expressed by the equations

$$0 = (DivT)^j = T^{ij}_{;i}$$
$$= T^{ij}_{,i} + \Gamma^i_{ik}T^{kj} + \Gamma^j_{ik}T^{ik},$$

which, in the case of spherical symmetry, can be written as the system of two equations:

$$0 = T^{00}_{,0} + T^{01}_{,1} + \Gamma^i_{ik}T^{k0} + \Gamma^0_{ik}T^{ik} \tag{3.3.1}$$
$$0 = T^{01}_{,0} + T^{11}_{,1} + \Gamma^i_{ik}T^{k1} + \Gamma^1_{ik}T^{ik}. \tag{3.3.2}$$

Substituting the expressions (3.2.4) for the connection coefficients (1.1.10) into (3.3.1) and (3.3.2), gives the equivalent system

$$0 = T^{00}_{,0} + T^{01}_{,1} + \frac{1}{2}\left(\frac{2A_t}{A} + \frac{B_t}{B}\right)T^{00} + \frac{1}{2}\left(\frac{3A'}{A} + \frac{B'}{B} + \frac{4}{r}\right)T^{01}$$
$$+ \frac{B_t}{2A}T^{11} \tag{3.3.3}$$

$$0 = T^{01}_{,0} + T^{11}_{,1} + \frac{1}{2}\left(\frac{A_t}{A} + \frac{3B_t}{B}\right)T^{01} + \frac{1}{2}\left(\frac{A'}{A} + \frac{2B'}{B} + \frac{4}{r}\right)T^{11}$$
$$+ \frac{A'}{2B}T^{00} - 2\frac{r}{B}T^{22}. \tag{3.3.4}$$

Now if one could use equations to eliminate the derivative terms A_t, A', B_t and B' in (3.3.3) and (3.3.4) in favor of of expressions involving the undifferentiated unknowns A, B and T, then system (3.3.3), (3.3.4) would take the form of a system of conservation laws with source terms. Indeed, T^{00} and T^{01} serve as the conserved quantities, T^{10} and T^{11} are the fluxes, and what is left, written as a function of the undifferentiated variables (A, B, T^{00}, T^{01}), would play the role of a source term. (For example, in a fractional step scheme designed to simulate the initial value problem, the variables A and B could be "updated" to time $t_j + \Delta t$ by the supplemental equations (1.3.2) and (1.3.4) or (1.3.3) and (1.3.4) after the conservation law step is implemented using the known values of A and B at time t_j. We carry this out in detail in the next chapter.) The system then closes once one writes T^{11} as a function of (A, B, T^{00}, T^{01}). There is a problem here, however. Equations (1.3.2)-(1.3.4) can be used to eliminate the terms A_r, B_t and B_r, but (3.3.3) and (3.3.4) also contain terms involving A_t, a quantity that is not given in the initial data and is not directly evolved by equations (1.3.2)-(1.3.4). The way to resolve this is to incorporate the A_t term into the conserved quantities. For general equations involving A_t, this is not possible. A natural change of T variables that eliminates the A_t terms from (3.3.3), (3.3.4), is to write the equations in terms of the values that T takes in flat Minkowski space. That is, define T_M in terms of T, by

$$T^{00} = \frac{1}{A} T_M^{00},$$

$$T^{01} = \frac{1}{\sqrt{AB}} T_M^{01}, \tag{3.3.5}$$

$$T^{11} = \frac{1}{B} T_M^{11},$$

where the subscript denotes Minkowski, c.f. (1.3.9)-(1.3.11). It then follows that T_M is given by

$$T_M^{00} = \left\{ (p + \rho c^2) \frac{c^2}{c^2 - v^2} - p \right\},$$

$$T_M^{01} = (p + \rho c^2) \frac{cv}{c^2 - v^2}, \tag{3.3.6}$$

$$T_M^{11} = \left\{ (p + \rho c^2) \frac{v^2}{c^2 - v^2} + p \right\},$$

where v denotes the fluid speed as measured by an inertial observer fixed with respect to the radial coordinate r, c.f. (1.3.12)-(1.3.14). Substituting (3.3.5) into (3.3.3), (3.3.4), the A_t terms cancel out, and we obtain the system

$$0 = \{T_M^{00}\}_{,0} + \left\{ \sqrt{\frac{A}{B}} T_M^{01} \right\}_{,1} + \frac{1}{2} \frac{B_t}{B} (T_M^{00} + T_M^{11})$$

$$+ \frac{1}{2} \sqrt{\frac{A}{B}} \left(\frac{A'}{A} + \frac{B'}{B} + \frac{4}{r} \right) T_M^{01} \tag{3.3.7}$$

$$0 = \{T_M^{01}\}_{,0} + \left\{ \sqrt{\frac{A}{B}} T_M^{11} \right\}_{,1} + \frac{1}{2} \sqrt{\frac{A}{B}} \left\{ 2 \frac{B_t}{\sqrt{AB}} T_M^{01} + \left(\frac{B'}{B} + \frac{4}{r} \right) T_M^{11} \right.$$

$$\left. + \frac{A'}{A} T_M^{00} - 4r T^{22} \right\}. \tag{3.3.8}$$

The following proposition states that system (3.3.7), (3.3.8) is equivalent, (in the weak sense), to the original system $DivT = 0$.

Proposition 5 *If A and B are given Lipschitz continuous functions defined on the domain $r \geq r_0$, $t \geq 0$, then T_M is a weak solution of (3.3.7) and (3.3.8) if and only if T is a weak solution of $DivT = 0$ in this domain.*

Proof: For simplicity, and without loss of generality, take the weak formulation with test functions compactly supported in $r > r_0$, $t > 0$, so that the boundary integrals do not appear in the weak formulations. (Managing the boundary integrals is straightforward.) The variables T_M^{ij} solve (3.3.7) weakly if

$$
0 = \int_{r_0}^{\infty} \int_0^{\infty} \left\{ -T^{00}\varphi_t - \sqrt{\frac{A}{B}} T^{01} \varphi_r \right.
$$
$$
\left. + \left[\frac{1}{2}\frac{B_t}{B}\left(T^{00} + T^{11}\right) + \frac{1}{2}\sqrt{\frac{A}{B}}\left(\frac{A'}{A} + \frac{B'}{B} + \frac{4}{r}\right)T^{01}\right]\varphi \right\} dr\, dt
$$
$$
= \int_{r_0}^{\infty} \int_0^{\infty} \left\{ -T_M^{00} A\varphi_t - T_M^{01} A\varphi_r \right. \tag{3.3.9}
$$
$$
\left. + \left[\frac{1}{2}\frac{B_t}{B}\left(AT_M^{00} + BT_M^{11}\right) + \frac{1}{2}A\left(\frac{A'}{A} + \frac{B'}{B} + \frac{4}{r}\right)T_M^{01}\right]\varphi \right\} dr\, dt.
$$

Set $\psi = A\varphi$, whereby $A\varphi_t = \psi_t - \frac{A_t}{A}\psi$. Using this change of test function, (3.3.9) becomes

$$
0 = \int_{r_0}^{\infty} \int_0^{\infty} \left\{ -T^{00}\psi_t + T^{00}\frac{A_t}{A}\psi - T^{01}\psi' + T^{01}\frac{A'}{A}\psi \right.
$$
$$
\left. + \left[\frac{1}{2}\frac{B_t}{B}\left(T^{00} + \frac{B}{A}T^{11}\right) + \frac{1}{2}\left(\frac{A'}{A} + \frac{B'}{B} + \frac{4}{r}\right)T^{01}\right]\psi \right\} dr\, dt.
$$
$$
= \int_{r_0}^{\infty} \int_0^{\infty} \left\{ -T^{00}\psi_t - T^{01}\psi' + \left[\frac{1}{2}\left(\frac{2A_t}{A} + \frac{B_t}{B}\right)T^{00} \right.\right.
$$
$$
\left.\left. + \frac{1}{2}\left(\frac{3A'}{A} + \frac{B'}{B} + \frac{4}{r}\right)T^{01} + \frac{B_t}{2A}T^{11}\right]\psi \right\} dr\, dt, \tag{3.3.10}
$$

which is the weak formulation of (3.3.3). We deduce that T_M solves (3.3.7) for every Lipschitz continuous test function φ if and only if T solves (3.3.10), (the weak form of $T_{;i}^{0i} = 0$), for all Lipschitz continuous test functions ψ. That weak solutions of (3.3.8) are weak solutions of $T_{;i}^{1i} = 0$ follows by a similar argument. $\square$

It is now possible to use equations (1.3.2)-(1.3.4) as identities to substitute for derivatives of metric components A and B, thereby eliminating the corresponding derivatives of A and B from the source terms of equations (3.3.7),

(3.3.8). Doing this, we obtain the system of equations (1.4.1), (1.4.2), which was our goal. However, depending on the choice of equation to drop, either (1.3.2) or (1.3.3), it is not clear that if we use the dropped equation to substitute for derivatives in (3.3.7), (3.3.8), that the resulting system of equations will imply that $DivT = 0$ continues to hold, the assumption we based the substitution on in the first place. The following theorem states that (1.4.1), (1.4.2) is equivalent to $DivT = 0$ in the weak sense:

Theorem 9 *Assume that A, B are Lipschitz continuous functions, and that $T \in L^\infty$, on the domain $r \geq r_0$, $t \geq 0$. Assume also that (1.3.2) holds at $t = 0$, and that (1.3.3) holds at $r = r_0$. Then A, B, T are weak solutions of (1.3.2), (1.3.3), (1.3.4) and $DivT = 0$ if and only if A, B, T_M are weak solutions of either system (1.3.2), (1.3.4), (1.4.1), (1.4.2), or system (1.3.3), (1.3.4), (1.4.1), (1.4.2).*

Proof: Without loss of generality, we consider the case when we drop equation (1.3.3), and use (1.3.2), (1.3.4) and $DivT = 0$ to evolve the metric, and we ask whether we can take the modified system (1.4.1) and (1.4.2) in place of $DivT = 0$. In this case, we must justify the use of (1.3.3) in eliminating the B_t terms in going from $DivT = 0$ to system (1.4.1) and (1.4.2). That is, it remains only to show that equations (1.3.2) and (1.3.4) together with system (1.4.1) and (1.4.2) imply that (1.3.3) holds, assuming (1.3.3) holds at $r = r_0$. (If so, then by substitution, it then follows that $DivT = 0$ also holds.)

Note that we can almost reconstruct (3.3.3), the first component of $DivT = 0$, by reverse substituting (1.3.2), (1.3.4) into (1.4.1). To see this, first note that we can add (1.3.2) and (1.3.4) to obtain

$$\frac{A'}{A} + \frac{B'}{B} - rB\kappa(T_M^{00} + T_M^{11}) = 0. \qquad (3.3.11)$$

Equation (3.3.11) is an identity that we may add to (1.4.1) to obtain

$$0 = \{T_M^{00}\}_{,0} + \left\{\sqrt{\frac{A}{B}}T_M^{01}\right\}_{,1} - \frac{1}{2}r\sqrt{AB}\kappa\left(T_M^{00} + T_M^{11}\right)T_M^{01}$$
$$+ \frac{1}{2}\sqrt{\frac{A}{B}}\left(\frac{A'}{A} + \frac{B'}{B} + \frac{4}{r}\right)T_M^{01}. \qquad (3.3.12)$$

Adding and subtracting

$$\frac{1}{2}\frac{B_t}{B}\left(T_M^{00} + T_M^{11}\right) \qquad (3.3.13)$$

to the RHS of (3.3.12) and using

$$H^{01} = -\frac{B_t}{rB} - \sqrt{AB}\kappa T_M^{01}, \qquad (3.3.14)$$

(c.f. (1.3.3) and (3.2.6)), we have

$$0 = \{T_M^{00}\}_{,0} + \left\{\sqrt{\frac{A}{B}} T_M^{01}\right\}_{,1} + \frac{1}{2}\sqrt{\frac{A}{B}}\left(\frac{A'}{A} + \frac{B'}{B} + \frac{4}{r}\right) T_M^{01}$$

$$+ \frac{1}{2}\frac{B_t}{B}\left(T_M^{00} + T_M^{11}\right) + \frac{1}{2}r\left(T_M^{00} + T_M^{11}\right)H^{01}. \tag{3.3.15}$$

Note that all but the last term on the RHS of (3.3.15) is equal to the first component of $DivT$, and so

$$T^{0i}_{;i} = -\frac{1}{2}r\left(T_M^{00} + T_M^{11}\right)H^{01}.$$

Therefore, if A, B, and T_M are solutions to (1.3.2), (1.3.4), (3.3.15), and (3.3.4), it follows that

$$H^{i0}_{;i} = G^{i0}_{;i} - \kappa T^{i0}_{;i}$$
$$= \kappa\frac{rB^2T^{11}}{2}H^{01}, \tag{3.3.16}$$

because $G^{i0}_{;i} = 0$ is an identity. But $H^{00} \equiv 0$ holds because we assume (1.3.2), and hence (3.3.16) implies that

$$H^{01}_{,1} + fH^{01} = 0,$$

where $f \equiv \Gamma^i_{i1} + 2\Gamma^1_{01} - \kappa\frac{rB^2T^{11}}{2} \in L^\infty$. Since we assume that $H^{01} = 0$ on the boundary $r = r_0$, it follows from Corollary 1 that $H^{01} \equiv 0$. □

It remains to identify conditions under which T_M^{11} is a function of (T_M^{00}, T_M^{01}) assuming that T has the form of a stress tensor for a perfect fluid, (3.3.6). A calculation shows that, in this case, the following simplifications occur:

$$T_M^{00} - T_M^{11} = \rho c^2 - p, \tag{3.3.17}$$
$$T_M^{00}T_M^{11} - (T_M^{01})^2 = p\rho c^2. \tag{3.3.18}$$

Using (3.3.17) and (3.3.18) we see that only the first terms on the RHS of (1.4.1), (1.4.2) depend on v, and the only term that is not linear in ρ and p is the third term on the RHS of (1.4.2). We state and prove the following theorem:

Theorem 10 *Assume that $0 < p < \rho c^2$, $0 < \frac{dp}{d\rho} < c^2$. Then T_M^{11} is a function of T_M^{00} and T_M^{01} so long as (ρ, v) lie in the domain $D = \{(\rho, v) : 0 < \rho, |v| < c\}$.*

Proof: We may write (3.3.17) and (3.3.18) in the form

$$T_M^{00} - T_M^{11} = f_1(\rho), \tag{3.3.19}$$
$$T_M^{00}T_M^{11} - (T_M^{01})^2 = f_2(\rho). \tag{3.3.20}$$

Since $\frac{df_1}{d\rho} = c^2 - p' \geq c^2 - \sigma^2 > 0$, it follows that the function f_1 is one-to-one with respect to ρ. Also, $\frac{df_2}{d\rho} = p'\rho c^2 + pc^2 \geq pc^2 > 0$, so the function f_2 is also one-to-one in ρ. Consequently, the function $h = f_2 \cdot f_1^{-1}$ is one-to-one, and thus

$$T_M^{00} T_M^{11} - (T_M^{01})^2 = h(T_M^{00} - T_M^{11}). \tag{3.3.21}$$

Now introduce the linear and invertible change of variables
$x = T_M^{00} - T_M^{11}$, $y = T_M^{01}$, $z = T_M^{11}$, whereby (3.3.21) becomes

$$(x+z)z - y^2 = h(x). \tag{3.3.22}$$

Equation (3.3.22) is quadratic in z, and so we may solve it directly, obtaining

$$z = \frac{-x \pm \sqrt{x^2 + 4(y^2 + h(x))}}{2}. \tag{3.3.23}$$

From (3.3.23), we conclude that for any (x, y), there are *two* values of z, though only one of these will correspond to values of ρ and v in the domain D. That is, since

$$x = T_M^{00} - T_M^{11} = \rho c^2 - p > 0, \tag{3.3.24}$$

and $z = T_M^{11} > 0$, it follows that there is at most one solution of (3.3.23) in the domain D, namely

$$z = \frac{-x + \sqrt{x^2 + 4(y^2 + h(x))}}{2}. \tag{3.3.25}$$

We conclude that if (ρ, v) lies in the domain D, then for each value of T_M^{00} and T_M^{01}, there exists precisely one value of T_M^{11}. $\square$

A calculation shows that in the case $p = \sigma^2 \rho$, $\sigma = constant$, the formula for T_M^{11} in terms of (T_M^{00}, T_M^{01}) is given by

$$T_M^{11} = \frac{1 + 2K_*}{2K_*} \left\{ T_M^{00} - \sqrt{(T_M^{00})^2 - \frac{4K_*}{(1 + 2K_*)^2} \left(K_*(T_M^{00})^2 + (T_M^{01})^2 \right)} \right\} \tag{3.3.26}$$

where

$$K_* = \frac{\sigma^2 c^2}{(c^2 - \sigma^2)^2}. \tag{3.3.27}$$

3.4 Statement of the General Problem

Our results concerning the weak formulation of the Einstein equations (1.3.2)-(1.3.5) assuming spherical symmetry given in Theorem 9 can be summarized as follows. Assume that A, B are Lipschitz continuous functions, and that $T \in L^\infty$, on the domain $r \geq r_0$, $t \geq 0$. Then (1.3.2)-(1.3.5) are equivalent to two different systems which take the form of a system of conservation laws with source terms. In the first case, we have shown that weak solutions of the system (1.3.2), (1.3.4) together with equations (3.3.7), (3.3.8) (for $DivT = 0$), will solve (1.3.2)-(1.3.5) weakly, so long as (1.3.3) holds at $r = r_0$. This reduces the Einstein equations with spherical symmetry to a system of equations of the general form

$$u_t + f(u, A, B)_x = \bar{g}(u, A, B, A', B_t, B', x), \tag{3.4.1}$$
$$A_x = h^0(u, A, B, x), \tag{3.4.2}$$
$$B_x = h^1(u, A, B, x), \tag{3.4.3}$$

where $u = (T_M^{00}, T_M^{01})$ agree with the conserved quantities that appear in the conservation law $divT_M = 0$ in flat Minkowski space. (Here "prime" denotes $\frac{\partial}{\partial x}$ since we are using x in place of r.) It is then valid to use equations (1.3.2)-(1.3.4) to eliminate all derivatives of A and B from the RHS of system (3.4.1), by which we obtain the system (1.3.2), (1.3.4), (1.4.1), (1.4.2), a system that closes to make a nonlinear system of conservation laws with source terms, taking the general form

$$u_t + f(u, A, B)_x = g(u, A, B, x),$$
$$A_x = h^0(u, A, B, x), \tag{3.4.4}$$
$$B_x = h^1(u, A, B, x),$$

which reproduces (1.4.3),(1.4.4) of Chapter 1. Weak solutions of (3.4.4) will satisfy (1.3.3) so long as (1.3.3) is satisfied on the boundary $r = r_0$.

In the second case, we have shown that weak solutions of the system (1.3.3), (1.3.4) together with equations (3.3.7), (3.3.8) (for $DivT = 0$), will solve (1.3.2)-(1.3.5) weakly, so long as (1.3.2) holds at $t = 0$. This reduces the Einstein equations with spherical symmetry to an alternative system of equations of the general form

$$u_t + f(u, A, B)_x = \bar{g}(u, A, B, A', B_t, B', x), \tag{3.4.5}$$
$$A_x = h^0(u, A, B, x), \tag{3.4.6}$$
$$B_t = h_*^1(u, A, B, x). \tag{3.4.7}$$

It is then valid to use equations (1.3.2)-(1.3.4) to eliminate all derivatives of A and B from the RHS of system (3.4.5), by which we obtain the system (1.3.3), (1.3.4), (1.4.1), (1.4.2), a system that closes to make a nonlinear system of conservation laws with source terms, taking the general form

$$u_t + f(u, A, B)_x = g(u, A, B, x), \tag{3.4.8}$$
$$A_x = h^0(u, A, B, x), \tag{3.4.9}$$
$$B_t = h^1_*(u, A, B, x). \tag{3.4.10}$$

Weak solutions of (3.4.8) will satisfy (1.3.2) so long as (1.3.2) is satisfied at $t = 0$.

3.5 Wave Speeds

In this section we conclude by calculating the wave speeds associated with system (1.4.1)-(1.4.2). Because A and B enter as undifferentiated source terms, it follows from (1.4.1)-(1.4.2) that for spherically symmetric flow, the only wave speeds in the problem will be the characteristic speeds for the fluid. Loosely speaking, the gravitational field is "dragged along" passively by the fluid when spherical symmetry is imposed. From this we conclude that there is no lightlike propagation, (that is, no gravity waves), in spherical symmetry, even when there is matter present. (This is the conclusion of Birkoff's theorem for the empty space equations, [35].)

The easiest way to calculate the wave speeds for the fluid is from the Rankine-Hugoniot jump conditions in the limit as the shock strength tends to zero. To start, note that the components of the 4-velocity $\mathbf{w}$ for a spherically symmetric fluid (1.2.4) are $w^0 = \frac{dt}{ds}$, $w^1 = \frac{dr}{ds}$, $w^2 = w^3 = 0$. Since $-1 = g(\mathbf{w}, \mathbf{w})$, the components w^0 and w^1 are not independent, and in particular, $-1 = -(w^0)^2 A + (w^1)^2 B$. We define fluid speed v as the speed measured by an observer fixed in (t, r) coordinates. That is, the speed is the change in distance per change in time as measured in an orthonormal frame with timelike vector parallel to ∂_t and spacelike vector parallel to ∂_r. It follows that the speed is given by $v = x/a$, where

$$\mathbf{w} = a\frac{\partial_t}{\sqrt{-g_{00}}} + x\frac{\partial_r}{\sqrt{g_{11}}}. \tag{3.5.1}$$

Taking the inner product of $\mathbf{w}$ with ∂_t and then with ∂_r, we find that $a = w^0\sqrt{-g_{00}}$ and $x = w^1\sqrt{g_{11}}$, and hence

$$v = \frac{w^1}{w^0}\sqrt{\frac{B}{A}}, \tag{3.5.2}$$

whereby,

$$(w^0)^2 = \frac{1}{A(c^2 - v^2)}. \tag{3.5.3}$$

Using (3.5.2) and (3.5.3) in (1.2.4), it follows that the components of the energy-momentum tensor take the following simplified form, which is valid globally in the (t, r) coordinate system:

$$T^{00} = \frac{1}{A}\left\{(p + \rho c^2)\frac{c^2}{c^2 - v^2} - p\right\}$$

$$T^{01} = \frac{1}{\sqrt{AB}}(p + \rho c^2)\frac{cv}{c^2 - v^2}$$

$$T^{11} = \frac{1}{B}\left\{(p + \rho c^2)\frac{v^2}{c^2 - v^2} + p\right\}.$$

Note that these components are equal to the components of the stress tensor in flat Minkowski space, times factors involving A and B that account for the fact that the spacetime is not flat. Using (1.3.9)-(1.3.11) we can write the Rankine-Hugoniot jump conditions in the form

$$s[T_M^{00}] = \sqrt{\frac{A}{B}}[T_M^{01}], \tag{3.5.4}$$

$$s[T_M^{01}] = \sqrt{\frac{A}{B}}[T_M^{11}]. \tag{3.5.5}$$

From (3.5.4)-(3.5.5), we deduce that wave speeds for the system (1.4.1)-(1.4.2) are $\sqrt{A/B}$ times the wave speeds in the Minkowski metric case, and this holds globally throughout the (t, r) coordinate system. (See [27].) Eliminating s from (3.5.4) and (3.5.5), yields

$$[T_M^{01}]^2 = [T_M^{00}][T_M^{11}]. \tag{3.5.6}$$

Now take the left fluid state on a shock curve to be (ρ_L, v_L), and the right fluid state to be (ρ, v). For a spherically symmetric perfect fluid, (3.5.6) defines the right velocity v as a function of the right density ρ. Then to obtain the fluid wave speeds, just substitute this function into (3.5.4), solve for s, and take the limit as $\rho \to \rho_L$. Following this procedure, (3.5.6) simplifies to

$$\frac{(v - v_L)^2}{(c^2 - v^2)(c^2 - v_L^2)} = \frac{[p][\rho]}{(p + \rho c^2)(p_L + \rho_L c^2)}. \tag{3.5.7}$$

Note that equation (3.5.7) can be written as a quadratic in v, and hence there are two solutions. The '+' solutions will yield the 2-shocks, and the '-' the 1-shocks. Dividing both sides of (3.5.7) by $(\rho - \rho_L)^2$ and taking the limit as $\rho \to \rho_L$, we see that

$$\frac{dp}{d\rho} = \frac{(p + c^2\rho)^2}{(c^2 - v^2)^2}\left(\frac{dv}{d\rho}\right)^2. \tag{3.5.8}$$

Solving (3.5.5) for s we obtain,

$$s = \sqrt{\frac{A}{B}}\frac{\left[(p + \rho c^2)\frac{v^2}{c^2 - v^2} + p\right]}{\left[(p + \rho c^2)\frac{cv}{c^2 - v^2}\right]}, \tag{3.5.9}$$

and taking the limit as $\rho \to \rho_L$, we obtain

$$\lambda_\pm = \sqrt{\frac{A}{B} \frac{\left[(p' + c^2)\frac{v^2}{c^2 - v^2} + (p + \rho c^2)\frac{2vv'(c^2 - v^2) + 2v^3 v'}{(c^2 - v^2)^2} + p'\right]}{\left[(p' + c^2)\frac{cv}{c^2 - v^2} + (p + \rho c^2)\frac{cv'(c^2 - v^2) + 2cv^2 v'}{(c^2 - v^2)^2}\right]}},$$

$$= \sqrt{\frac{A}{B} \frac{\left[(p' + c^2)\frac{v^2}{c^2 - v^2} + (p + \rho c^2)\frac{2c^2 vv'}{(c^2 - v^2)^2} + p'\right]}{\left[(p' + c^2)\frac{cv}{c^2 - v^2} + (p + \rho c^2)\frac{cv'(c^2 + v^2)}{(c^2 - v^2)^2}\right]}}.$$

(Here the plus/minus on RHS is determined by the two possible signs of $v' = dv/d\rho$ as allowed by (3.5.8).) After substituting for $dv/d\rho$ using (3.5.8), and simplifying, we obtain

$$\lambda_\pm = \sqrt{\frac{A}{B} \frac{\left[(p' + c^2)\frac{v^2}{c^2 - v^2} \pm \frac{2c^2 v\sqrt{p'}}{(c^2 - v^2)} + p'\right]}{\left[(p' + c^2)\frac{cv}{c^2 - v^2} \pm \frac{c(c^2 + v^2)\sqrt{p'}}{(c^2 - v^2)}\right]}},$$

$$= \sqrt{\frac{A}{B} \frac{\left[(p' + c^2)v^2 \pm 2c^2 v\sqrt{p'} + p'(c^2 - v^2)\right]}{\left[(p' + c^2)cv \pm c(c^2 + v^2)\sqrt{p'}\right]}},$$

$$= c\sqrt{\frac{A}{B} \frac{\left[v^2 \pm 2v\sqrt{p'} + p'\right]}{\left[vp' \pm (c^2 + v^2)\sqrt{p'} + c^2 v\right]}},$$

$$= c\sqrt{\frac{A}{B} \frac{\left[v \pm \sqrt{p'}\right]^2}{\left[v \pm \sqrt{p'}\right]\left[c^2 \pm v\sqrt{p'}\right]}}.$$

This gives the wave speeds as:

$$\lambda_\pm = c\sqrt{\frac{A}{B}} \frac{\sqrt{p'} \pm v}{v\sqrt{p'} \pm c^2}. \tag{3.5.10}$$

(For example, the formula for λ_- results from choosing '-' in (3.5.7).) The following theorem demonstrates that the system (1.4.1)-(1.4.2) is strictly hyperbolic whenever the particles are moving at less than the speed of light:

Proposition 6 *Assume that*

$$|v| < c,$$

so that the particle trajectory has a timelike tangent vector. Then wave speeds for the general relativistic Euler equations (1.4.1)-(1.4.2) satisfy $\lambda_- < \lambda_+$.

Proof: To determine where the wave speeds are equal, set λ_- equal to λ_+ and solve for v to obtain $v^2 = c^2$. Next, substitute $v = 0$ into λ_- and λ_+ to verify that $\lambda_- < \lambda_+$ when $v^2 < c^2 A/B$. Proposition 6 follows directly. $\square$

As a final comment, we note that Proposition 6 is true because it is true in a locally inertial coordinate system centered at any point P in spacetime. Indeed, in such a coordinate system, the connection coefficients vanish at P, and the metric components match those of the Minkowski metric to first order in a neighborhood of P. As a result, the general relativistic Euler equations reduce to the classical relativistic Euler equations at P. Since it is known in Special Relativity that the Euler equations are strictly hyperbolic for timelike particles, [27], it follows that the same must be true in General Relativity. Other pointwise properties, such as genuine nonlinearity and the Lax entropy inequalities, [26, 18], can be verified for the spherically symmetric general relativistic equations in a similar manner.

4

Existence and Consistency for the Initial Value Problem

4.1 Introduction

In this chapter, taken from [13], we present a proof that shock wave solutions of (1.3.2)-(1.3.5), (1.2.4) and (1.3.1), defined outside a ball of fixed total mass, exist up until some positive time $T > 0$, and we prove that the total mass $M_\infty = \lim_{r \to \infty} M(r, t)$ is constant throughout the time interval $[0, T)$. A local existence theorem is all that we can expect for system (1.3.2)-(1.3.5) in general because black holes are singularities in standard Schwarzschild coordinates, $B = \frac{1}{1 - \frac{2M}{r}} \to \infty$ at a black hole, and black holes can form in finite time. For these solutions, the fluid variables ρ, p and $\mathbf{w}$, and the components of the stress tensor T^{ij}, are *discontinuous*, and the metric components A and B are *Lipschitz continuous*, at the shock waves, c.f. (1.3.2) and (1.3.4). Since (1.3.5) involves second derivatives of A and B, it follows that these solutions satisfy (1.3.2)-(1.3.5) only in the weak sense of the theory of distributions. Thus our theorem establishes the consistency of the initial value problem for the Einstein equations at the weaker level of shock waves.

To be precise, assume the initial boundary conditions

$$\rho(r, 0) = \rho_0(r), \quad v(r, 0) = v_0(r), \quad for \ r > r_0,$$

$$M(r_0, t) = M_{r_0}, \quad v(r_0, t) = 0, \quad for \ t \geq 0,$$

$$(4.1.1)$$

where r_0 and M_{r_0} are positive constants, and assume that the no black hole and finite total mass conditions,

$$\frac{2M(r, t)}{r} < 1, \quad \lim_{r \to \infty} M(r, t) = M_\infty < \infty, \qquad (4.1.2)$$

hold at $t = 0$. For convenience, assume further that

$$\lim_{r \to \infty} r^2 T_M^{00}(r,t) = 0, \tag{4.1.3}$$

holds at $t = 0$, c.f., (1.3.15), (4.1.2). The main result of this chapter can be stated as follows:

Theorem 11 *Assume that the initial boundary data satisfy (4.1.1)-(4.1.3), and assume that there exist positive constants L, V and $\bar{v}$ such that the initial velocity and density profiles $v_0(r)$ and $\rho_0(r)$ satisfy*

$$TV_{[r,r+L]} \ln \rho_0(\cdot) < V, \quad TV_{[r,r+L]} \ln \left(\frac{c + v_0(\cdot)}{c - v_0(\cdot)} \right) < V, \quad |v_0(r)| < \bar{v} < c, \tag{4.1.4}$$

for all $r_0 \le r < \infty$, where $TV_{[a,b]} f(\cdot)$ denotes the total variation of the function f over the interval $[a,b]$. Then a bounded weak (shock wave) solution of (1.3.2)-(1.3.5), satisfying (4.1.1) and (4.1.2), exists up to some positive time $T > 0$. Moreover, the metric functions A and B are Lipschitz continuous functions of (r,t), and (4.1.4) continues to hold for $t < T$ with adjusted values for V and $\bar{v}$ that are determined from the analysis.

Note that the theorem allows for arbitrary numbers of interacting shock waves, of arbitrary strength. Note that by (1.3.2), (1.3.4), the metric components A and B will be no smoother than Lipschitz continuous when shocks are present, and thus since (1.3.5) is second order in the metric, it follows that (1.3.5) is only satisfied in the weak sense of the theory of distributions. Note finally that $\lim_{r \to \infty} M(r,t) = M_\infty$ is a *non-local* condition.

In the argument leading to (3.4.4) of Chapter 3, we showed that when the metric components A and B are Lipschitz continuous, system (1.3.2)-(1.3.5) is weakly equivalent to the following system of equations obtained by replacing (1.3.3) and (1.3.5) with the 0- and 1-components of $DivT = 0$,

$$\{T_M^{00}\}_{,0} + \left\{ \sqrt{\frac{A}{B}} T_M^{01} \right\}_{,1} = -\frac{2}{x} \sqrt{\frac{A}{B}} T_M^{01}, \tag{4.1.5}$$

$$\{T_M^{01}\}_{,0} + \left\{ \sqrt{\frac{A}{B}} T_M^{11} \right\}_{,1} = -\frac{1}{2} \sqrt{\frac{A}{B}} \left\{ \frac{4}{x} T_M^{11} + \frac{(B-1)}{x}(T_M^{00} - T_M^{11}) \right. \tag{4.1.6}$$

$$\left. + 2\kappa x B(T_M^{00} T_M^{11} - (T_M^{01})^2) - 4x T^{22} \right\},$$

$$\frac{B'}{B} = -\frac{(B-1)}{x} + \kappa x B T_M^{00}, \tag{4.1.7}$$

$$\frac{A'}{A} = \frac{(B-1)}{x} + \kappa x B T_M^{11}. \tag{4.1.8}$$

This is the system of conservation laws with source terms which we have written in the compact form (1.4.3), (1.4.4),

$$u_t + f(\mathbf{A}, u)_x = g(\mathbf{A}, u, x),$$
$$\mathbf{A}' = h(\mathbf{A}, u, x).$$

Here

$$u = (T_M^{00}, T_M^{01}) \equiv (u^0, u^1),$$
$$\mathbf{A} = (A, B),$$
$$f(\mathbf{A}, u) = \sqrt{\frac{A}{B}} \left(T_M^{01}, T_M^{11} \right), \qquad (4.1.9)$$

and it follows from (4.1.5)-(4.1.8) that

$$g(\mathbf{A}, u, x) = \left(g^0(\mathbf{A}, u, x), g^1(\mathbf{A}, u, x) \right), \qquad (4.1.10)$$

$$h(\mathbf{A}, u, x) = \left(h^0(\mathbf{A}, u, x), h^1(\mathbf{A}, u, x) \right), \qquad (4.1.11)$$

where

$$g^0(\mathbf{A}, u, x) = -\frac{2}{x} \sqrt{\frac{A}{B}} T_M^{01}, \qquad (4.1.12)$$

$$g^1(\mathbf{A}, u, x) = -\frac{1}{2} \sqrt{\frac{A}{B}} \left\{ \frac{4}{x} T_M^{11} + \frac{(B-1)}{x} (T_M^{00} - T_M^{11}) \right. \qquad (4.1.13)$$
$$\left. + 2\kappa x B (T_M^{00} T_M^{11} - (T_M^{01})^2) - 4x T^{22} \right\},$$

and

$$h^0(\mathbf{A}, u, x) = \frac{(B-1)A}{x} + \kappa x A B T_M^{11}, \qquad (4.1.14)$$

$$h^1(\mathbf{A}, u, x) = -\frac{(B-1)B}{x} + \kappa x B^2 T_M^{00}. \qquad (4.1.15)$$

The vector $h(\mathbf{A}, u, x)$ was just obtained by solving (1.3.2) and (1.3.4) for A' and B'. Note that we have set $x \equiv x^1 \equiv r$, and will use x in place of r in the analysis to follow since this is standard notation in the literature on hyperbolic conservation laws. Note also that we write t when we really mean ct, in the sense that t must be replaced by ct whenever we put dimensions of time, i.e., factors of c, into our formulas. We interpret this as taking $c = 1$ when convenient.

Recall that a new twist in formulation (1.4.3), (1.4.4) is that the conserved quantities are taken to be the the energy and momentum densities

$u = (u^0, u^1) = (T_M^{00}, T_M^{01})$ of the relativistic compressible Euler equations in flat Minkowski spacetime–quantities that, unlike the entries of T, are independent of the metric. Recall also that, (remarkably), all time derivatives of metric components cancel out from the equations when this change of variables is made, c.f. Section 3.3 and [12]. We take advantage of this formulation in the numerical method which we develop here for the study of the initial value problem.

The proof of Theorem 1 is based on a fractional step Glimm scheme, c.f. [19, 11]. The fractional step method employs a Riemann problem step that simulates the source free conservation law $u_t + f(\mathbf{A}, u)_x = 0$, $(\mathbf{A} \equiv Const)$, followed by an ODE step that accounts for the sources present in both f and g. The idea for the numerical scheme is to stagger discontinuities in the metric with discontinuities in the fluid variables so that the conservation law step as well as the ODE step of the method are both generated in grid rectangles on which the metric components $\mathbf{A} = (A, B)$, (as well as x), are constant. At the end of each timestep, we solve $A' = h(\mathbf{A}, u, x)$ and re-discretize, to update the metric sources. Part of our proof involves showing that the ODE step $u_t = g(\mathbf{A}, u, x) - \nabla_\mathbf{A} f \cdot \mathbf{A}'$, with h substituted for $\mathbf{A}'$, accounts for both the source term g, as well as the *effective sources* that are due to the discontinuities in the metric components at the boundaries of the grid rectangles.

Because of our formulation (1.4.3), (1.4.4), only the flux f in the conservation law step, depends on $\mathbf{A}$. From this dependence we conclude that the only effect of the metric on the Riemann problem step of the method is to change the wave speeds, but not the states of the waves that solve the Riemann problem. Thus, on the Riemann problem step, when we assume $p = \sigma^2 \rho$, we can apply the estimates derived in Chapter 2 for flat Minkowski spacetime $\mathbf{A} = (1, 1)$. Applying these results, it follows that the Riemann problem is *globally* solvable in each grid cell, and the total variation in $\ln \rho$ is non-increasing on the Riemann problem step of our fractional step scheme, (Lemma 7). Thus we need only estimate the increase in total variation of $\ln \rho$ for the ODE step of the method, in order to obtain a local total variation bound, and hence compactness of the numerical approximations up to some time $T > 0$.

One nice feature of our method is that the ODE that accomplishes the ODE step of the method, turns out to have surprisingly nice properties. Indeed, a phase portrait analysis shows that $\rho > 0$, $|v| < c$ is an invariant region for solution trajectories. (Since x and $\mathbf{A}$ are taken to be constant on the ODE step, the ODE's form an autonomous system at each grid cell.) We also show that even though the ODE's are quadratic in ρ, solutions of the ODE's do not blow up, but in fact remain bounded for all time. It follows that the fractional step scheme is defined and bounded so long as the Courant-Freidrichs-Levy (CFL) condition is maintained, [26]. We show that the CFL bound depends only on the supnorm of the metric component $\|B\|_\infty$, together with the supnorm $\|S\|_\infty$, where $S \equiv S(x, t) = x\rho(x, t)$. We go on to prove that all norms in the problem are bounded by a function that depends only on $\|B\|_\infty \|S\|_\infty$, and $\|TV_L \ln \rho(\cdot, t)\|_\infty$, where the latter denotes the sup of the total variation over intervals of L. By this we show that the solution can be extended up until the first time at which one of these three norms tends to infinity. (Our analysis rules out the possibility that $v \to c$ before one of these norms blows up.) The condition $B \to \infty$ corresponds to the formation of a black hole, and $\rho \to \infty$ corresponds to the formation of a naked singularity, (because the scalar curvature satisfies $R = \{c^2 - 3\sigma^2\}\rho$). It is known that black holes can form in solutions of the Einstein equations, and it is an open problem whether or not naked singularities can form, or whether we can have $\|S\|_\infty \to \infty$, or $\|TV_L \ln \rho(\cdot, t)\|_\infty \to \infty$, in some other way.

The main technical problem is to prove that the total mass $M_\infty = \frac{\kappa}{2} \int_{r_0}^\infty \rho r^2 \, dr$ is bounded. The problem is that, in our estimates, the growth of ρ depends on M and the growth of M depends on ρ, and M is defined by a *non-local* integral. Thus, an error estimate of order Δx for $\Delta \rho$ after one time step, is not sufficient to bound the total mass M_∞ after one time step.

In Section 4.2, we introduce the notation and state the main result, Theorem 13. In Section 4.3 we define the fractional step scheme, the so called *locally inertial Glimm scheme*. The fractional step scheme involves a Riemann problem step to handle the hyperbolic part and an ODE step to account for geometric and other sources in each grid cell. In Section 4.4 we analyze the Riemann problem step of the fractional step method, and in Sections 4.5 and 4.6 we analyze the ODE solutions in each grid cell. Estimates for the synthesis of the the Riemann step followed by the ODE step under iteration in approximate solutions generated by the fractional step scheme are derived in Section 4.7. Uniform bounds on the total variation of approximate solutions at each fixed time are obtained in Sections 4.8. The method is to first assume the relevant norms in the problem are bounded, and then to obtain estimates derived in terms of these bounds that are strong enough to eliminate the starting assumptions. The convergence of the residual, (the measure of how far the approximate solutions are from true weak solutions), is proven in Section 4.9. This involves cancellation of errors from the ODE step with errors from the Riemann step, and justifies the incorporation of terms from the flux into the ODE step. Concluding remarks are made in the final Section 4.10.

4.2 Preliminaries

The starting point of our analysis is the following theorem, which is a restatement of Theorem 2, Chapter 3. This theorem implies the equivalence of system (1.4.3),(1.4.4) with the Einstein equations (1.3.2)-(1.3.5) for weak, (shock wave), solutions, so long as (1.3.3) is treated as a constraint that holds so long as it holds at the boundary $x = r_0$. (Again, we use the variable x in place of r in order to conform with standard notation, c.f. [26]).

Theorem 12 *Let $u(x,t), \mathbf{A}(x,t)$ be weak solutions of (1.4.3),(1.4.4) in the domain*

$$D \equiv \{(x,t) : r_0 \le x < \infty, 0 \le t < T\}, \tag{4.2.1}$$

for some $r_0 > 0$, $T > 0$. Assume that u is in $L_{loc}^\infty(D)$, and that $\mathbf{A}$ is locally Lipschitz continuous in D, by which we mean that for any open ball B centered at a point in D, there is a constant $C > 0$ such that

$$|\mathbf{A}(x_2,t_2) - \mathbf{A}(x_1,t_1)| \le C\{|x_2 - x_1| + |t_2 - t_1|\}. \tag{4.2.2}$$

Then u and $\mathbf{A}$ satisfy all four Einstein equations (1.3.2)- (1.3.5) throughout D if and only if the equation (1.3.3),

$$-\frac{\dot{B}}{xB} = \kappa ABT^{01},$$

holds at the boundary $x = r_0$. In this case, it follows that the equivalent forms (1.3.7), (1.3.8) of (1.3.2),(1.3.3), respectively, also hold in the strong sense throughout D.

Note that for our problem, the constraint (1.3.8), and therefore (1.3.3), is implied by the boundary conditions

$$M(r_0,t) = M_{r_0}, \tag{4.2.3}$$
$$v(r_0,t) = 0, \tag{4.2.4}$$

alone, because, using (1.3.13), equation (1.3.8) translates into

$$\dot{M} = -\frac{\kappa}{2}\sqrt{\frac{A}{B}}\frac{c^2 + \sigma^2}{c^2 - v^2}cv\rho x^2,$$

which, in light of (4.2.3), (4.2.4), is an identity at the boundary $x = r_0$.

It follows from Theorem 12 that in order to establish Theorem 11, it suffices only to prove the corresponding existence theorem for system (1.4.3)-(1.4.4) in domain D. The equation (1.3.3) will then follow as an identity on weak solutions because it is met at the boundary. It follows that if we construct weak solutions for which v is uniformly bounded and for which ρ decreases fast enough, then we can apply (1.3.3) as $x \to \infty$ to conclude that

$$\lim_{x \to \infty} \dot{M}(x,t) = 0. \tag{4.2.5}$$

This is our strategy for proving that the total mass is finite.

Before stating the main theorem precisely, a few preliminary comments regarding system (1.4.3)-(1.4.4) are in order. First note that system (1.4.3)-(1.4.4) closes once we express T_M^{11} and T^{22} on the RHS of (4.1.9), (4.1.10) and (4.1.11), as a function of the conserved quantities $u = (u^0, u^1) \equiv (T_M^{00}, T_M^{01})$. From (1.2.4) it follows that

$$T^{22} = \frac{p}{x^2} = \frac{\sigma^2 \rho}{x^2}, \tag{4.2.6}$$

and this can be expressed in terms of u via the mapping (4.2.21) discussed below. To write T_M^{11} as a function of u, use the identities, (c.f. (4.69),(4,70) of [12]),

$$T_M^{00} - T_M^{11} = \rho c^2 - p \equiv f_1(\rho), \tag{4.2.7}$$
$$T_M^{00} T_M^{11} - (T_M^{01})^2 = p\rho c^2 \equiv f_2(\rho). \tag{4.2.8}$$

By (4.2.7),

$$\rho = f_1^{-1}(T_M^{00} - T_M^{11}), \tag{4.2.9}$$

and using this in (4.2.8), one can in general solve (4.2.8) for T_M^{11}. In the case $p = \sigma^2 \rho$, a calculation gives

$$T_M^{11} = \frac{2\zeta + 1}{2\zeta} \left\{ 1 - \sqrt{1 - \frac{4\zeta}{(2\zeta + 1)^2} \left(\zeta + \left[\frac{T_M^{01}}{T_M^{00}} \right]^2 \right)} \right\} T_M^{00}, \tag{4.2.10}$$

where

$$\zeta = \frac{\sigma^2 c^2}{(c^2 - \sigma^2)^2}. \tag{4.2.11}$$

It is readily verified that the quantity under the square root sign is positive so long as

$$\left[\frac{T_M^{01}}{T_M^{00}} \right] < 1 + \frac{1}{2\zeta},$$

which holds in light of (1.3.17). It follows that (4.2.10) defines T_M^{11} as a smooth, single valued function of the conserved quantities $(u^0, u^1) \equiv (T_M^{00}, T_M^{01})$. Other than its existence, we will not need the explicit formula for T_M^{11} given in (4.2.10).

We are free to analyze the state space for system (1.4.3)-(1.4.4) in the plane of conserved quantities $u = (u^0, u^1) \equiv (T_M^{00}, T_M^{01})$, in the (ρ, u) plane, or in the plane of Riemann invariants (r, s) which are defined in terms of ρ and v via the special relativistic Euler equations in flat Minkowski spacetime, (assume $p = \sigma^2 \rho$, c.f. (2.5.73), (2.5.74) of Chapter 2),

$$r = \frac{1}{2} \ln \frac{c + v}{c - v} - \frac{K_0}{2} \ln \rho, \tag{4.2.12}$$
$$s = \frac{1}{2} \ln \frac{c + v}{c - v} + \frac{K_0}{2} \ln \rho, \tag{4.2.13}$$

where

$$K_0 = \frac{\sigma c}{c^2 + \sigma^2}. \tag{4.2.14}$$

(We use Roman "r" for the Riemann invariant to distinguish it from the radial variable "r", c.f. (2.5.73), (2.5.74).) In this section it is more convenient for us to use the variables

$$z \equiv s - r = K_0 \ln \rho, \tag{4.2.15}$$

$$w \equiv s + r = \ln \frac{c + v}{c - v}, \tag{4.2.16}$$

and we let $\mathbf{z}$ denote the vector

$$\mathbf{z} = (z, w) \equiv \left(K_0 \ln \rho, \ln \frac{c - v}{c + v} \right). \tag{4.2.17}$$

Given this, we use the following notation: As usual, the double norm $\| \cdot \|$ applied to a vector denotes Euclidean norm, so e.g., $\|u\| \equiv \sqrt{(u^0)^2 + (u^1)^2}$ and $\|\mathbf{z}\| \equiv \sqrt{(z)^2 + (w)^2)}$, and the single norm $| \cdot |$, when applied to scalars, denotes the the regular absolute value. But we use the special notation that $| \cdot |$, when applied to a vector, denotes the change in the z-component across the vector, so that, e.g.,

$$|\mathbf{z}| \equiv |z|. \tag{4.2.18}$$

Similarly, if γ denotes a wave with left state $\mathbf{z}_L$ and right state $\mathbf{z}_L$, (see (4.3.10) and (4.4.13)-(4.4.15) below), then we let

$$\|\gamma\| \equiv \sqrt{|z_R - z_L|^2 + |w_R - w_R|^2}, \tag{4.2.19}$$

$$|\gamma| \equiv |z_R - z_L|, \tag{4.2.20}$$

and we refer to $|\gamma|$ as the strength of the wave γ, c.f. [27, 19].

Equations (1.3.12), (1.3.13), and (4.2.12)-(4.2.16), define the mappings $\Psi : (\rho, v) \to (u^0, u^1)$ and $\Phi : (\rho, v) \to (z, w)$,

$$\begin{pmatrix} u^0 \\ u^1 \end{pmatrix} = \Psi \begin{pmatrix} \rho \\ v \end{pmatrix} \equiv \begin{pmatrix} \frac{c^4 + \sigma^2 v^2}{c^2 - v^2} \rho c^2 \\ \frac{(c^2 + \sigma^2) c v}{c^2 - v^2} \rho \end{pmatrix}, \tag{4.2.21}$$

$$\begin{pmatrix} z \\ w \end{pmatrix} = \Phi \begin{pmatrix} \rho \\ v \end{pmatrix} \equiv \begin{pmatrix} K_0 \ln \rho \\ \ln \frac{c + v}{c - v} \end{pmatrix}. \tag{4.2.22}$$

The following proposition states that the mappings Ψ and Φ define one to one regular maps between the respective domains:

Proposition 7 *The mapping*

$$\Phi : D \to R \qquad (4.2.23)$$

defined by (4.2.21) is smooth, one-to-one and onto, from domain

$$\mathcal{D} = \{(\rho, v) : 0 < \rho < \infty, |v| < c\}, \qquad (4.2.24)$$

to range

$$R = \{(u^0, u^1) : 0 < u^0 < \infty, |u^1| < \infty\}; \qquad (4.2.25)$$

and the mapping

$$\Phi : \mathcal{D} \to \hat{R}, \qquad (4.2.26)$$

defined in (4.2.22), is smooth, one-to-one and onto from domain $\mathcal{D}$ to

$$\hat{R} = \{(z, w) : -\infty < z < +\infty, -\infty < w < +\infty\}. \qquad (4.2.27)$$

Proof: This follows directly from (4.2.21) and (4.2.22).

The goal of this chapter is to prove the following theorem:

Theorem 13 *Let*

$$u_0(x) \equiv (u_0^0(x), u_0^1(x)) = \Psi(\rho_0(x), v_0(x)) = \Psi \circ \Phi^{-1}(z_0(x), w_0(x)),$$

and $\mathbf{A}_0(x) = (A_0(x), B_0(x))$ denote initial data for system (1.4.3),(1.4.4), defined for $x \geq r_0$. Assume that there exists positive constants V, L, and $\bar{v}$, such that

$$TV_{[x,x+L]} \ln \rho_0(\cdot) < V, \qquad (4.2.28)$$

$$TV_{[x,x+L]} \ln \frac{c + v_0(\cdot)}{c - v_0(\cdot)} < V, \qquad (4.2.29)$$

$$|v_0(x)| < \bar{v}, \qquad (4.2.30)$$

for all $x \geq r_0$. Assume that $B_0(x) = \frac{1}{1 - \frac{2M_0(x)}{x}}$, where the initial mass function $M_0(x)$ is given by

$$M_0(x) = M_{r_0} + \frac{\kappa}{2} \int_{r_0}^x u_0^0(r) r^2 \, dr, \qquad (4.2.31)$$

(c.f. (1.3.15)), and assume that M_0 satisfies the conditions

$$\lim_{x \to \infty} M_0(x) = M_\infty < \infty, \qquad (4.2.32)$$

and

$$1 - \frac{2M_0(x)}{x} = B_0^{-1}(x) > \bar{B}^{-1} > 0, \qquad (4.2.33)$$

respectively, for some fixed positive constants $M_{r_0} < M_\infty$, and $\bar{B} < \infty$. Assume finally that

$$A_0(x) = A_{r_0} \exp \int_{r_0}^x \left\{ \frac{B_0(r) - 1}{r} + \kappa r B_0(r) T_M^{11}(u_0(r)) \right\} dr \qquad (4.2.34)$$

for some fixed positive constant $A_{r_0} > 0$, so that

$$A_0(r_0) = A_{r_0} > 0. \qquad (4.2.35)$$

Given this, we conclude that there exists $T > 0$, and functions $u(x,t), \mathbf{A}(x,t)$ defined on $x \geq r_0$, $0 \leq t < T$, such that $u(x,t), \mathbf{A}(x,t)$ is a weak solution of system (4.1.5),(4.1.6), (1.3.2)-(1.3.4), together with the initial-boundary conditions

$$\rho(x,0) = \rho_0(x), \quad v(x,0) = v_0(x), \qquad (4.2.36)$$

$$\mathbf{A}(r_0,t) = \left(A_{r_0}, \frac{1}{1 - \frac{2M_{r_0}}{r_0}} \right), \qquad (4.2.37)$$

$$v(r_0,t) = 0. \qquad (4.2.38)$$

Moreover, the solution $u, \mathbf{A}$ satisfies the following:
(i) For each $t \in [0,T)$ there exists a constant $V(t) < \infty$ such that

$$TV_{[x,x+L]} \ln \rho(\cdot,t') < V(t), \qquad (4.2.39)$$

$$TV_{[x,x+L]} \ln \frac{c + v(\cdot,t')}{c - v(\cdot,t')} < V(t), \qquad (4.2.40)$$

for all $t' \leq t$.
(ii) For each $x \geq r_0$ and $t \in [0,T)$,

$$0 < A(x,t), B(x,t) < \infty, \qquad (4.2.41)$$

and

$$\lim_{x \to \infty} M(x,t) = M_\infty. \qquad (4.2.42)$$

(iii) For each closed bounded set $\mathcal{U} \subset \{(x,t) : x \geq r_0, \ 0 \leq t < T\}$, there exists a constant $C(\mathcal{U}) < \infty$ such that,

$$\|\mathbf{A}(x_2,t_2) - \mathbf{A}(x_1,t_1)\| < C(\mathcal{U}) \{|x_2 - x_1| + |t_2 - t_1|\}, \qquad (4.2.43)$$

and

$$\int_{r_0}^x \|u(r,t_2) - u(r,t_1)\| dr < C(\mathcal{U})|t_2 - t_1|. \qquad (4.2.44)$$

Here, (4.2.39) and (4.2.40) imply that the functions $z(\cdot, t)$ and $w(\cdot, t)$ are functions of locally bounded total variation at each fixed time $t < T$, and the bounds are uniform over bounded sets in $x \geq r_0$, $0 \leq t < T$. Estimates (4.2.39) and (4.2.40) also imply that $\rho > 0$ and $|v| < c$, and therefore that $u^0 > 0$ throughout $x \geq r_0$, $0 \leq t < T$. The inequality (4.2.41) says that $B = \frac{1}{1 - \frac{2M}{x}} > 0$, and hence that $\frac{2M}{x} < 1$ for $t < T$, the condition that no black holes have formed before time T. Inequality (4.2.43) says that the metric components A and B are locally Lipschitz continuous functions in $x \geq r_0$, $0 \leq t < T$, and (4.2.43) says that $u(x, t)$ is L^1-Lipschitz continuous in time, uniformly on bounded sets. Note that (4.2.31), (4.2.34) are included to guarantee that equations (1.3.2) and (1.3.4), (and so also (1.4.4)), are satisfied at time $t = 0$.

4.3 The Fractional Step Scheme.

In this section we define the approximate solutions $u_{\Delta x}$, $\mathbf{A}_{\Delta x} = (A_{\Delta x}, B_{\Delta x})$ of system (1.4.3), (1.4.4) constructed by a fractional step Glimm scheme. Again, we have set $x \equiv x^1 \equiv r$, and we write t in place of ct, in the sense that t must be replaced by ct whenever we put dimensions of time, (that is, factors of c), into our formulas.

Let $\Delta x \ll 1$ denote a mesh length for space and Δt a mesh length for time, and assume that

$$\frac{\Delta x}{\Delta t} = \Lambda, \tag{4.3.1}$$

so that Λ^{-1} is the Courant number. We choose

$$\Lambda \geq Max \left\{ 2\sqrt{\frac{A}{B}} \right\}, \tag{4.3.2}$$

where the maximum is taken over all values that appear in the approximate solution. This guarantees the Courant-Friedrichs-Levy (CFL) condition, the condition that the mesh speed be greater than the maximum wave speed in the problem. (That is, $\sqrt{\frac{A}{B}}$ is the speed of light in standard Schwarzschild coordinates, and the factor of two accounts for the fact that waves emanate from the center of the mesh rectangles in our approximation scheme. Of course, as part of our proof, we must show that the maximum on the RHS of (4.3.2) exists.) Let (x_i, t_j) be mesh points in an unstaggered grid defined on the domain

$$D = \{r_0 \leq x \leq \infty, t \geq 0\}, \tag{4.3.3}$$

by setting

$$x_i = r_0 + i\Delta r,$$
$$t_j = j\Delta t.$$

Each mesh point (x_i, t_j), $i \geq 0$, $j \geq 0$, is positioned at the bottom center of the grid rectangle $\mathcal{R}_{ij}$,

$$\mathcal{R}_{ij} = \{x_{i-\frac{1}{2}} \leq x < x_{i+\frac{1}{2}}, \ t_j \leq t < t_{j+1}\}, \tag{4.3.4}$$

where $x_{i+\frac{1}{2}} = (i \pm \frac{1}{2})\Delta x$. Let $\mathcal{R}_{i_0 j}$ denote the half rectangle $\{x_{i_0} \leq x < x_{i_0+\frac{1}{2}}, \ t_j \leq t < t_{j+1}\}$ at the boundary $x = r_0$. (This is diagrammed in Figure 3.)

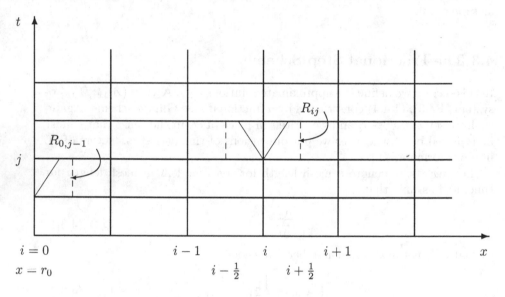

The mesh rectangles $\mathcal{R}_{ij}$.

Figure 3

In the approximation scheme, the metric source $\mathbf{A} = (A, B)$ is approximated by the constant value $\mathbf{A}_{ij}$ in each grid rectangle $\mathcal{R}_{ij}$, so set

$$\mathbf{A}_{\Delta x}(x, t) = \mathbf{A}_{ij} \text{ for } (x, t) \in \mathcal{R}_{ij}, \tag{4.3.5}$$

for values of $\mathbf{A}_{ij}$ to be defined presently. It follows that $\mathbf{A}_{\Delta x}$ is discontinuous along each line $x = x_{i+\frac{1}{2}}$, $i = 0, ..., \infty$, and at each time $t = t_j$. In our definition below, values of $\mathbf{A}_{ij}$ are determined from values $\mathbf{A}_{i,j-1}$ and $u_{\Delta x}$ at time t_j-, by solving (1.4.4), using the boundary condition $\mathbf{A} = \mathbf{A}_{r_0} = \left(A_{r_0}, \frac{1}{1-\frac{2Mr_0}{r_0}}\right)$ at the boundary $x = r_0$.

We now define $u_{\Delta x}$ by induction. First assume that $u_{\Delta x}$ is given by piecewise constant states u_{ij} at time $t = t_j+$ as follows:

$$u_{\Delta x}(x, t) = u_{ij} \text{ for } x_i \leq x < x_{i+1}, \ t = t_j + . \tag{4.3.6}$$

This poses the Riemann problem

$$u_0(x) = \begin{cases} u_L = u_{i-1,j} & x < x_i, \\ u_R = u_{ij} & x > x_i, \end{cases} \tag{4.3.7}$$

for the system

$$u_t + f(\mathbf{A}_{ij}, u)_x = 0, \tag{4.3.8}$$

at the bottom center of each mesh rectangle $\mathcal{R}_{ij}$, $i \geq 1$. When $i = 0$, the boundary condition $v = 0$ at $x_0 = r_0$ replaces the left state, and so in this case, the piecewise constant state u_{0j} at time $t = t_j+$ poses the boundary Riemann problem

$$u_0(x) = \begin{cases} v = 0 & x = r_0, \\ u_R = u_{0,j} & x > r_0, \end{cases} \tag{4.3.9}$$

Let $u_{ij}^{RP}(x, t)$ denote the solution of (4.3.6), (4.3.7) for $(x, t) \in \mathcal{R}_{ij}$, and let

$$u_{\Delta x}^{RP}(x, t) = u_{ij}^{RP}(x, t) \text{ for } (x, t) \in \mathcal{R}_{ij}. \tag{4.3.10}$$

Equation (4.3.10) defines the Riemann problem step of the fractional step scheme. Note that since $\mathbf{A}_{\Delta x} = \mathbf{A}_{ij}$ is constant in each $\mathcal{R}_{ij}$, it follows that system (4.3.8) is just the special relativistic Euler equations for $p = \sigma^2 \rho$, with a rescaled flux. We discuss the solution of this Riemann problem in detail in Section 4.4. We conclude there that the solution $u_{ij}^{RP}(x, t)$ consists of a 1-wave γ_{ij}^1 followed by a 2-wave γ_{ij}^2 for all $i > 0$, it consists of a single 2-wave $\gamma_{0j}^2 = 0$ at the boundary $i = 0$, and the waves γ_{ij}^p all have sub-luminous speeds so long as (4.3.2) holds. It follows that (4.3.2) guarantees that the waves in the Riemann problem (4.3.7), (4.3.7), never leave $\mathcal{R}_{ij}$ in one time step, c.f. Proposition 9 below.

The Riemann problem step of the method ignores the effect of the source term g in system (1.4.3), and also ignores the effect of the discontinuities in the flux $f(\mathbf{A}, u)$ due to discontinuities in $\mathbf{A}$ at the boundaries $x_{i-\frac{1}{2}}$ of $\mathcal{R}_{ij}$. These effects are accounted for in the ODE step. For the ODE step of the fractional step scheme, the idea is to use the Riemann problem solutions as initial data, and solve the ODE's

$$u_t = G(\mathbf{A}, u, x) \equiv g - \mathbf{A}' \cdot \nabla_\mathbf{A} f, \tag{4.3.11}$$

for one time step, thus defining the approximate solution in $\mathcal{R}_{ij}$. The first term on the RHS of (4.3.11) accounts for the sources on the RHS of (1.4.3), and the second term accounts for the discontinuities in $\mathbf{A}$ at the boundaries $x_{i-\frac{1}{2}}$. Now by (4.1.13),

$$g \equiv g(\mathbf{A}, u, x) = \left(g^0(\mathbf{A}, u, x), g^1(\mathbf{A}, u, x)\right),$$

where

$$g^0(\mathbf{A}, u, x) = -\frac{2}{x}\sqrt{\frac{A}{B}} T_M^{01}, \tag{4.3.12}$$

$$g^1(\mathbf{A}, u, x) = -\frac{1}{2}\sqrt{\frac{A}{B}} \left\{ \frac{4}{x}T_M^{11} + \frac{(B-1)}{x}(T_M^{00} - T_M^{11}) \right. \tag{4.3.13}$$

$$\left. +2\kappa x B(T_M^{00}T_M^{11} - (T_M^{01})^2) - 4xT^{22} \right\}.$$

By (4.1.9),

$$\nabla_{\mathbf{A}} f \equiv \nabla_{\mathbf{A}} f(\mathbf{A}, u) \equiv \left(\nabla_{\mathbf{A}} f^0, \nabla_{\mathbf{A}} f^1\right) = \left(\frac{1}{2}\frac{1}{\sqrt{AB}}T_M^{01}, -\frac{1}{2}\frac{1}{B\sqrt{AB}}T_M^{11}\right), \tag{4.3.14}$$

and by (4.1.15),

$$\mathbf{A}' = h \equiv \left(h^0(\mathbf{A}, u, x)), h^1(\mathbf{A}, u, x))\right), \tag{4.3.15}$$

where

$$h^0(\mathbf{A}, u, x) = \frac{(B-1)A}{x} + \kappa x A B T_M^{11},$$

$$h^1(\mathbf{A}, u, x) = -\frac{(B-1)B}{x} + \kappa x B^2 T_M^{00}. \tag{4.3.16}$$

It follows from (4.3.14)-(4.3.16) that

$$\mathbf{A}' \cdot \nabla_{\mathbf{A}} f(\mathbf{A}, u, x) = \frac{1}{2}\sqrt{\frac{A}{B}}\delta\left(T_M^{01}, T_M^{11}\right), \tag{4.3.17}$$

where

$$\delta = \frac{A'}{A} - \frac{B'}{B} = \frac{2(B-1)}{x} - \kappa x B\left(T_M^{00} - T_M^{11}\right). \tag{4.3.18}$$

Using (4.3.12), (4.3.13) and (4.3.18) and simplfying, we find that the ODE step should be

$$u_t = G, \tag{4.3.19}$$

where

$$G^0(\mathbf{A}, u, x) = -\frac{1}{2}\sqrt{\frac{A}{B}} T_M^{01} \left\{ \frac{2(B+1)}{x} - \kappa x B \left(T_M^{00} - T_M^{11} \right) \right\}, \tag{4.3.20}$$

$$G^1(\mathbf{A}, u, x) = -\frac{1}{2}\sqrt{\frac{A}{B}} \left\{ \frac{4}{x} T_M^{11} + \frac{B-1}{x} \left(T_M^{00} + T_M^{11} \right) + \tag{4.3.21} \right.$$

$$\left. \kappa x B \left[T_M^{00} T_M^{11} - 2 \left(T_M^{01} \right)^2 + \left(T_M^{11} \right)^2 \right] - 4x T^{22} \right\}.$$

Since $u = (T_M^{00}, T_M^{01})$, and T_M^{11}, T^{22} are given as functions of u in (4.2.6), (4.2.10), respectively, it follows that the right hand sides of (4.3.20) and (4.3.21) determine well defined functions of $(\mathbf{A}, u, x)$. It follows that G, as defined in (4.3.20), (4.3.21) also satisfies

$$G(\mathbf{A}, u, x) = g(\mathbf{A}, u, x) - \mathbf{A}' \cdot \nabla_{\mathbf{A}} f(\mathbf{A}, u, x),$$

where (4.3.12), (4.3.13) and (4.3.17) define g and $\mathbf{A}' \cdot f$ as functions of $(\mathbf{A}, u, x)$.

We can now define the ODE step of the method. Let $\hat{u}(t, u_0)$ denote the solution to the initial value problem

$$\hat{u}_t = G(\mathbf{A}_{ij}, \hat{u}, x) = g(\mathbf{A}_{ij}, \hat{u}, x) - \mathbf{A}' \cdot \nabla_{\mathbf{A}} f(\mathbf{A}_{ij}, \hat{u}, x),$$
$$\hat{u}(0) = u_0, \tag{4.3.22}$$

where $G(\mathbf{A}, \hat{u}, x)$ is defined in (4.3.20), (4.3.21), and $g(\mathbf{A}, u, x)$ and $\mathbf{A}' \cdot f(\mathbf{A}, u, x)$ are defined in (4.3.12), (4.3.13) and (4.3.17), respectively. It follows that

$$\hat{u}(t, u_0) - u_0 = \int_0^t \hat{u}_t \, dt$$
$$= \int_0^t \left\{ g(\mathbf{A}_{ij}, \hat{u}(\xi, u_0), x) - \mathbf{A}' \cdot \nabla_{\mathbf{A}} f(\mathbf{A}_{ij}, \hat{u}(\xi, u_0), x) \right\} d\xi.$$

Define the approximate solution $u_{\Delta x}(x, t)$ on each mesh rectangle $\mathcal{R}_{ij}$ by the formula

$$u_{\Delta x}(x, t) = u_{\Delta x}^{RP}(x, t) + \int_{t_j}^t \left\{ G(\mathbf{A}_{ij}, \hat{u}(\xi - t_j, u_{\Delta x}^{RP}(x, t)), x) \right\} d\xi \tag{4.3.23}$$

$$= u_{\Delta x}^{RP}(x, t) + \int_{t_j}^t \left\{ g(\mathbf{A}_{ij}, \hat{u}(\xi - t_j, u_{\Delta x}^{RP}(x, t)), x) \right\} d\xi$$

$$- \int_{t_j}^t \left\{ \mathbf{A}' \cdot \nabla_{\mathbf{A}} f(\mathbf{A}_{ij}, \hat{u}(\xi - t_j, u_{\Delta x}^{RP}(x, t)), x) \right\} d\xi.$$

Thus on each mesh rectangle $\mathcal{R}_{ij}$, $u_{\Delta x}(x,t)$ is equal to $u_{\Delta x}^{RP}(x,t)$ plus a correction that defines the ODE step of the method.

To complete the definition of $u_{\Delta x}$ by induction, it remains only to define the constant states $A_{i,j+1}$ on $\mathcal{R}_{i,j+1}$, and $u_{i,j+1} = u_{\Delta x}(x, t_{j+1}+)$ for $x_i \leq x < x_{i+1}$, in terms of the values of $u_{\Delta x}, \mathbf{A}_{\Delta x}$ defined for $t_j \leq t < t_{j+1}$. For this we use Glimm's method of random choice, c.f. [10, 26]. Thus let

$$\mathbf{a} \equiv \{a_j\}_{j=0}^{\infty} \in \mathbf{\Pi}, \tag{4.3.24}$$

denote a (fixed) random sequence, $0 < a_j < 1$, where $\mathbf{\Pi}$ denotes the infinite product measure space $\Pi_{i=0}^{\infty}(0,1)_j$, where $(0,1)_j$ denotes the unit interval $(0,1)$ endowed with Lebesgue measure, $0 < j < \infty$. (For convenience, assume WLOG that $a_0 = \frac{1}{2}$.) Then, assuming that $u_{\Delta x}, \mathbf{A}_{\Delta x}$ is defined up to time $t < t_{j+1}$, define

$$u_{i,j+1} = u_{\Delta x}(x_i + a_{j+1}\Delta x, t_{j+1}-), \tag{4.3.25}$$

$$M_{\Delta x}(x, t_{j+1}) = M_{r_0} + \frac{\kappa}{2} \int_{r_0}^{x} u_{\Delta x}^0(r, t_{j+1}-)r^2\, dr, \tag{4.3.26}$$

c.f. (1.3.15).[6] In terms of these, define the functions

$$B_{\Delta x}(x, t_{j+1}) = \frac{1}{1 - \frac{2M_{\Delta x}(x, t_{j+1})}{x}}, \tag{4.3.27}$$

and

$$A(x, t_{j+1}) = A_{r_0} \exp \int_{r_0}^{x} \left\{ \frac{B_{\Delta x}(r, t_{j+1}) - 1}{r} \right.$$
$$\left. + \kappa r B_{\Delta x}(r, t_{j+1}) T_M^{11}(u_{\Delta x}(r, t_{j+1})) \right\}\, dr, \tag{4.3.28}$$

c.f. (1.3.6) and (1.3.4). Finally, in terms of these, define

$$M_{i,j+1} = M(x_i, t_{j+1}), \tag{4.3.29}$$

$$B_{i,j+1} = B(x_i, t_{j+1}) = \frac{1}{1 - \frac{2M_{i,j+1}}{x_i}}, \tag{4.3.30}$$

and

$$A_{i,j+1} = A(x_i, t_{j+1}). \tag{4.3.31}$$

---

[6] By (4.3.25), the approximate solution depends on the choice of sample sequence **a**. In the last section, we prove that for almost every choice of sample sequence, a subsequence of approximate solutions converges to a weak solution of (1.4.4).

Let $\mathbf{A}_{i,j+1} = (A_{i,j+1}, B_{i,j+1})$ denote the constant value for $A_{\Delta x}$ on $\mathcal{R}_{i,j+1}$. This completes the definition of the approximate solution $u_{\Delta x}$ by induction. Note that (4.3.26)-(4.3.28) imply that when $\rho > 0$, $|v| < c$, we have

$$B_{\Delta x}(x, t_j) \geq 1, \tag{4.3.32}$$

$$B_{\Delta x}(r_0, t_j) = \frac{1}{1 - \frac{2M r_0}{r_0}} \equiv B_{r_0}, \tag{4.3.33}$$

$$A_{\Delta x}(x, t_j) \geq A_{r_0}, \tag{4.3.34}$$

for all $x \geq r_0, j \geq 0$. Note also that as a consequence of (4.3.26), (4.3.27) and (4.3.28), equations (1.3.2) and (1.3.4) hold in the form

$$\frac{B'_{\Delta x}(x,t_j)}{B} = -\frac{B_{\Delta x}(x,t_j) - 1}{x} + \kappa B_{\Delta x}(x,t_j) x T_M^{11}(u_{\Delta x}(x,t_j)), \tag{4.3.35}$$

$$\frac{A'_{\Delta x}(x,t_j)}{A} = +\frac{B_{\Delta x}(x,t_j) - 1}{x} + \kappa B_{\Delta x}(x,t_j) x T_M^{00}(u_{\Delta x}(x,t_j)). \tag{4.3.36}$$

Therefore,

$$\frac{\partial}{\partial x} \ln\{A_{\Delta x}(x,t_j) B_{\Delta x}(x,t_j)\} = \frac{A'}{A} + \frac{B'}{B}$$
$$\leq 4\kappa x B_{\Delta x}(x,t_j)(T_M^{00}(u_{\Delta x}(x,t_j)) + T_M^{11}(u_{\Delta x}(x,t_j))).$$

Integrating this from r_0 to x yields

$$A_{\Delta x}(x,t_j) B_{\Delta x}(x,t_j) \leq A_{r_0} B_{r_0} \exp\left\{\frac{8}{r_0} \int_{r_0}^x B_{\Delta x}(x,t_j) \frac{\kappa}{2} r^2 T_M^{00}(u_{\Delta x}(x,t_j))\right\}. \tag{4.3.37}$$

Inequalities (4.3.35)-(4.3.37) directly imply the following proposition:

Proposition 8 *Assume that there exist positive constants $\bar{M}$, $\bar{B}$, $\bar{S}$, $\bar{v}$, and integer $J > 0$, such that the approximate solution $u_{\Delta x}$, $\mathbf{A}_{\Delta x}$, defined as above, exists and satisfies*

$$M_{\Delta x}(x,t_j) \leq \bar{M}, \tag{4.3.38}$$

$$B_{\Delta x}(x,t_j) \leq \bar{B}, \tag{4.3.39}$$

$$0 \leq S_{\Delta x}(x,t_j) \equiv |x\rho_{\Delta x}(x,t_j)| \leq \bar{S} < \infty, \tag{4.3.40}$$

and

$$|v_{\Delta x}(x,t_j)| \le \bar{v} < c, \tag{4.3.41}$$

for all $x \ge r_0$, $j \le J$, so that by (1.3.12),

$$0 \le xu^0_{\Delta x}(x,t_j) \le \frac{c^2 + \sigma^2 \bar{v}^2}{c^2 - \bar{v}^2} \bar{S}. \tag{4.3.42}$$

Then

$$0 < \frac{A_{r_0}}{B_{\Delta x}(x,t_j)} \le \frac{A_{\Delta x}(x,t_j)}{B_{\Delta x}(x,t_j)} \le A_{\Delta x}(x,t_j) \le A_{\Delta x}(x,t_j)B_{\Delta x}(x,t_j)$$

$$\le A_{r_0}B_{r_0}exp\left\{\frac{8\bar{B}\bar{M}}{r_0}\right\} \equiv G_{AB}(\bar{B},\bar{M}), \tag{4.3.43}$$

and

$$|A'_{\Delta x}(x,t_j)| \le \left(\frac{1}{r_0} + \kappa\frac{c^2 + \sigma^2\bar{v}^2}{c^2 - \bar{v}^2}\bar{S}\right)G_{AB}(\bar{B},\bar{M}), \tag{4.3.44}$$

$$|B'_{\Delta x}(x,t_j)| \le \left(\frac{1}{r_0} + \kappa\frac{c^2 + \sigma^2\bar{v}^2}{c^2 - \bar{v}^2}\bar{S}\right)\bar{B}^2 \tag{4.3.45}$$

for all $x \ge r_0$, and $j \le J$.

Note that by (4.3.30), (4.3.31), (4.3.43)-(4.3.45) apply with $A_{\Delta x}(x,t_j)$, $B_{\Delta x}(x,t_j)$, replaced by A_{ij}, B_{ij}, respectively. Note also that (4.3.43) implies that

$$\Lambda = 2\sqrt{G_{AB}} \tag{4.3.46}$$

suffices to guarantee the CFL condition (4.3.2), and note that (4.3.44) and (4.3.45) imply

$$\left\|\frac{\Delta\mathbf{A}_{\Delta x}}{\Delta x}\right\| \le \left(\frac{1}{r_0} + \kappa\frac{c^2 + \sigma^2\bar{v}^2}{c^2 - \bar{v}^2}\bar{S}\right)(\bar{B}^2 + G_{AB}(\bar{B},\bar{M})), \tag{4.3.47}$$

where

$$\frac{\Delta\mathbf{A}_{\Delta x}}{\Delta x} = \frac{\mathbf{A}_{i+1,j} - \mathbf{A}_{ij}}{\Delta x}, \tag{4.3.48}$$

which gives the Lipschitz continuity in x of $A_{\Delta x}$ and $B_{\Delta x}$, respectively.

Proof: Inequality (4.3.43) follows directly form (4.3.37) in light of (4.7.30) and (4.3.26), and (4.3.44), (4.3.45) follow directly from (4.3.35), (4.3.36), and (4.3.43).

4.4 The Riemann Problem Step

In this section we discuss u_{ij}^{RP}, the solutions which constitute the Riemann problem step in the construction of $u_{\Delta x}$. For fixed (i, j), $u_{ij}^{RP}(x, t)$ is defined in (4.3.7), (4.3.8) as the solution of the Riemann problem

$$u_t + f(\mathbf{A}_{ij}, u)_x = 0, \tag{4.4.1}$$

$$u_0(x) = \left\{ \begin{array}{ll} u_L = u_{i-1,j} & x < 0 \\ u_R = u_{ij} & x \geq 0 \end{array} \right\}, \tag{4.4.2}$$

with the origin translated to the bottom center (x_i, t_j) of the mesh rectangle $\mathcal{R}_{ij} \equiv \{(x, t) : x_{i-\frac{1}{2}} < x \leq x_{i+\frac{1}{2}}, t_j \leq t < t_{j+1}\}$. Vector $\mathbf{A}_{ij}$ is constant on $\mathcal{R}_{ij}$. Assuming $p = \sigma^2 \rho$, system (4.4.1) takes the form

$$(T_M^{00})_{,t} + \left(\sqrt{\frac{A_{ij}}{B_{ij}}} T_M^{01} \right)_{,x} = 0, \tag{4.4.3}$$

$$(T_M^{01})_{,t} + \left(\sqrt{\frac{A_{ij}}{B_{ij}}} T_M^{11} \right)_{,x} = 0, \tag{4.4.4}$$

where T_M^{11} is given as a function of T_M^{00} and T_M^{01} in (4.2.10).

Proposition 9 *Assume that u_L and u_R correspond to values of ρ and v that lie in the region $\rho > 0$, $-c < v < c$. Then the Riemann problem (4.4.1), (4.4.2) has a unique solution consisting of elementary waves: shock waves and rarefaction waves. The solution is scale invariant, (is a function of x/t), and consists of a 1-wave γ_{ij}^1 followed by a 2-wave γ_{ij}^2. Moreover, the CFL condition (4.3.2) guarantees that the speeds of the waves are always smaller than the mesh speed $\frac{\Delta x}{\Delta t} = Max\left\{ 2\sqrt{\frac{A}{B}} \right\}$, and thus waves never interact during one time step.*

Proof: System (4.4.3)-(4.4.4) is the relativistic compressible Euler equations $DivT_M = 0$ in flat Minkowski spacetime, except for the constant factor $\sqrt{\frac{A_{ij}}{B_{ij}}}$ that multiplies the flux. Now the factor $\sqrt{\frac{A_{ij}}{B_{ij}}}$ changes the speeds of the waves, but does not affect the values of u on the elementary waves γ_{ij}^p. Indeed, the scale change $\bar{t} \to t/\sqrt{A_{ij}/B_{ij}}$ converts (4.4.1) into the Minkowski space problem $DivT_M = 0$, and so it follows from the frame invariance of the compressible Euler equations that (s, u_L, u_R) satisfies the Rankine-Hugoniot jump conditions

$$s[u] = [f] = \sqrt{\frac{A_{ij}}{B_{ij}}}[f_M],$$
(4.4.5)

for system (4.4.1), if and only if $(\bar{s}, u)$ satisfies the Minkowski jump conditions

$$\bar{s}[u] = [f_M],$$
(4.4.6)

where

$$s = \sqrt{\frac{A_{ij}}{B_{ij}}}\bar{s}.$$
(4.4.7)

(Recall that a shock with left state u_L, right state u_R, and speed s, is a weak solution of a conservation law $u_t + f(u)_x = 0$ if and only if the Rankine-Hugoniot jump relations $s[u] = [f]$ are satisfied.) Here f denotes the flux in (4.4.1), $f_M = f(1, 1, u)$ denotes the standard Minkowski flux, and $[\cdot]$ denotes the jump in a quantity from left to right across a shock. Thus the i-shock curves for system (4.4.1) agree with the i-shock curves for the system $u_t + f_M(u)_x) = 0$, when $\mathbf{A}_{ij} = (A_{ij}, B_{ij}) = (1, 1)$, [26]. Moreover, since $[u]$ tends to an eigen-direction and s tends to an eigenspeed as $[u] \to 0$ across a shock, it follows that the i-rarefaction curves $\mathbf{R}_i$ and i-shock curves $\mathbf{S}_i$ for system (4.4.1) are the same as the curves for the Minkowski system $u_t + f_M(u)_x = 0$, c.f. [26, 27, 10, 17, 7]. It follows that the factor $\sqrt{\frac{A_{ij}}{B_{ij}}}$ changes the speeds of the waves, but does not affect the values of u on the elementary waves γ_{ij}^p, as claimed.

It was shown in Chapter 2 that the Riemann problem for system $u_t + f_M(u)_x = 0$ has a unique solution consisting of a 1-wave followed by a 2-wave, and all wave speeds are subluminous so long as $\rho > 0$, $-c < v < c$. If we denote this solution by $[u_L, u_R]_M(x, t)$, then, (assuming $\rho > 0$, $-c < v < c$), it follows from (4.4.7) that the solution of (4.4.1), (4.4.2) is given by $[u_L, u_R](x, t) = [u_L, u_R]_M(x, \sqrt{\frac{A_{ij}}{B_{ij}}}t)$. Since, by Theorem (4), Chapter 2, all shock and characteristic speeds are sub-luminous for the Minkowski problem $DivT_M = 0$, $p = \sigma^2\rho$, it follows from (4.4.7) that wave speeds in the solution of the Riemann problem (4.4.1), (4.4.2) are bounded by $\sqrt{\frac{A}{B}}$, the speed of light in standard Schwarzschild coordinates. This verifies that if $\rho > 0$, $-c < v < c$, then the CFL condition (4.3.2) guarantees that all wave speeds in the solution u_{ij}^{RP} are bounded by the mesh speed $\frac{\Delta x}{\Delta t} = \Lambda$. Note that this implies that the constant states $u_{i-1,j}$, u_{ij} are maintained along the left and right boundaries of $\mathcal{R}_{ij}$ in the approximate solution u_{ij}^{RP}. $\square$

For fixed $\mathbf{A}_{ij} = (A_{ij}, B_{ij})$, let

$$[u_L, u_R] \equiv [u_L, u_R](x, t),$$
(4.4.8)

denote the solution of the Riemann problem (4.4.1), (4.4.2), and write

$$[u_L, u_R] = (\gamma^1, \gamma^2), \tag{4.4.9}$$

to indicate that the solution $[u_L, u_R](x, t)$ consists of the 1-wave γ^1 followed by the 2-wave γ^2. An elementary wave γ is itself a solution of a Riemann problem, in which case we write $[u_L, u_R] = \gamma$, and we call u_L and u_R the right and left states of the wave γ, respectively. In this case, define $|\gamma|$, the strength of the wave γ, by

$$|\gamma| = |K_0 \ln(u_L) - K_0 \ln(u_R)| = \left| K_0 \ln\left(\frac{u_L}{u_R}\right) \right|. \tag{4.4.10}$$

c.f. (4.2.14). (It is convenient in this section to incorporate K_0 into the wave strength so that the strength of the wave controls the jump in z across the wave, c.f. (2.7.94), Chapter 2.) For the general case $[u_L, u_R] = (\gamma^1, \gamma^2)$, we define the strength of the Riemann problem as the sum of the strengths of its elementary waves,

$$|[u_L, u_R]| = |\gamma^1| + |\gamma^2|. \tag{4.4.11}$$

The following proposition, special to the case $p = \sigma^2 \rho$, states that the sum of the strengths of elementary waves are non-increasing during wave interactions, so long as $\mathbf{A}_{ij}$ is constant.

Proposition 10 *Assume that $\mathbf{A}_{ij}$ is fixed. Let u_L, u_M, and u_R be any three states in the region $\rho > 0$, $-c < v < c$. Then*

$$|[u_L, u_R]| \leq |[u_L, u_M]| + |[u_M, u_R]|. \tag{4.4.12}$$

Proof: It was shown in Lemma 6 of Chapter 2, that (4.4.12) holds in the special relativistic case $div T_M = 0$. Since the effect of $\mathbf{A}_{ij}$ is to change the speeds of the elementary waves, but not the left and right states, in the solution of (4.4.1), (4.4.2), it follows that the estimate (4.4.12) continues to hold for arbitrary, (but constant), values of $\mathbf{A}_{ij}$. $\square$

Proposition 10 is a direct consequence of the geometry of shock and rarefaction curves derived in Chapter 2, and discussed further below, and is not true except in the special case $p = \sigma^2 \rho$, [27]. It follows from Proposition 10 that the only increases in the total variation of $\ln \rho_{\Delta x}(\cdot, t)$ in an approximate solution $u_{\Delta x}(\cdot, t)$ is due to increases that occur during the ODE steps (4.3.22). This is the basis for our analysis of convergence. Thus we analyze solutions in the $\mathbf{z}$-plane, $\mathbf{z} = (z, w) \equiv (K_0 \ln \rho, \ln \frac{c-v}{c+v})$, a $45°$ rotation of the plane of Riemann invariants (R, S), c.f.(4.2.12), (4.2.13).

Thus, let $\mathbf{z}_L$, $\mathbf{z}_R$ be the left and right states of a single elementary wave γ, and let γ denote both the *name* of the wave, as well as the *vector*

$$\gamma = \mathbf{z}_R - \mathbf{z}_L. \tag{4.4.13}$$

Let

$$\|\gamma\| = \|\mathbf{z}_R - \mathbf{z}_L\|, \tag{4.4.14}$$

and so we have

$$|\gamma| = |K_0 \ln \rho_R - K_0 \ln \rho_L| = |z_R - z_L| \le \|\gamma\|, \tag{4.4.15}$$

where K_0 is defined in (4.2.14). Note that because changes in $\mathbf{A}$ affect only the speeds of waves, it follows that $\gamma, |\gamma|$ and $\|\gamma\|$ depend only on $\mathbf{z}_L, \mathbf{z}_R$, and not on the value of $\mathbf{A_{ij}}$ used in the construction. We write

$$[\mathbf{z}_L, \mathbf{z}_R] \equiv [u_L, u_R] = (\gamma^1, \gamma^2), \tag{4.4.16}$$

to indicate that γ^1, γ^2 are the elementary 1- and 2-waves that solve the Riemann Problem with left state $\mathbf{z}_L = \Phi \circ \Psi^{-1} u_L$ and right state $\mathbf{z}_L = \Phi \circ \Psi^{-1} u_R$. We now summarize the results in Section 2.4, Chapter 2, regarding the geometry of shock and rarefaction curves as plotted in the $\mathbf{z}$-plane.

Let $S_i(\mathbf{z}_L)$ denote the i-shock curve emanating from the left state $\mathbf{z}_L$. That is, $\mathbf{z}_R \in S_i(\mathbf{z}_L)$ if and only if $[\mathbf{z}_L, \mathbf{z}_R]$ is a pure i-shock, [26]. It was shown in Section 2.4, Chapter 2, that all i-shock curves are translates of one another in the $\mathbf{z}$-plane, and 2-shock curves are just the reflection of the 1-shock curves about lines $z = const$. The following formula for the 1-shock curve is given in equations (2.5.75), (2.5.76) of Lemma 3, Chapter 2:

Lemma 9 *A state $\mathbf{z}_R$ lies on the 1-shock curve $S_1(\mathbf{z}_L)$ if and only if*

$$\Delta \mathrm{r} = -\frac{1}{2} \ln \{f_+(2K\zeta)\} - \frac{K_0}{2} \ln \{f_+(\zeta)\}, \tag{4.4.17}$$

$$\Delta \mathrm{s} = -\frac{1}{2} \ln \{f_+(2K\zeta)\} + \frac{K_0}{2} \ln \{f_+(\zeta)\}, \tag{4.4.18}$$

where

$$f_+(\zeta) = (1 + \zeta) + \sqrt{\zeta(1 + \zeta)}, \tag{4.4.19}$$

for some $0 \le \zeta < \infty$. Here

$$K = \frac{2\sigma^2 c^2}{(c^2 + \sigma^2)^2}, \tag{4.4.20}$$

and $\Delta \mathrm{r} = \mathrm{r}_R - \mathrm{r}_L$, $\Delta \mathrm{s} = \mathrm{s}_R - \mathrm{s}_L$, denote the change in the Riemann invariants across the shock.

Using (4.2.15),(4.2.16) we see that (4.4.17),(4.4.18) are equivalent to

$$\Delta w = -\ln\{f_+(2K\zeta)\}, \tag{4.4.21}$$

$$\Delta z = -K_0 \ln\{f_+(\zeta)\}. \tag{4.4.22}$$

Since (4.4.21),(4.4.22) describe the 1-shock curves for $0 \le \zeta < \infty$, it follows directly from these that 1-shock curves $S_1(\mathbf{z}_L)$ have a geometric shape in the $\mathbf{z}$-plane that is independent of $\mathbf{z}_L$. Thus, all 1-shock curves are translates of one another in the $\mathbf{z}$-plane, as claimed in Section 2.5, Chapter 2. We also showed in Chapter 2 that the 2-shock curve $S_2(\mathbf{z}_L)$ is the reflection of $S_1(\mathbf{z}_L)$ about the line $z = z_L$, (this follows directly from (2.5.77), (2.5.78) of Chapter 2.) From this, together with (4.4.21),(4.4.22), it follows that

$$|\Delta w| = \ln\{f_+(K_0^2\zeta)\}, \tag{4.4.23}$$

$$|\Delta z| = K_0 \ln\{f_+(\zeta)\}. \tag{4.4.24}$$

all along both the 1- and 2-shock curves. The next lemma implies the convexity of shock curves in the case $p = \sigma^2\rho$.

Lemma 10 *The shock equations (4.4.23), (4.4.24) imply that*

$$\sinh\left(\frac{|\Delta w|}{2}\right) = K_0 \sinh\left(\frac{|\Delta z|}{2K_0}\right), \tag{4.4.25}$$

from which it follows that (4.4.23), (4.4.24) define

$$|\Delta w| = H(|\Delta z|), \tag{4.4.26}$$

where the function H is given by

$$H(|\Delta z|) = \ln f_+\left(2K_0^2 \sinh^2\left\{\frac{|\Delta z|}{2K_0}\right\}\right) = 2\sinh^{-1}\left(K_0 \sinh\frac{|\Delta z|}{2K_0}\right). \tag{4.4.27}$$

The function H satisfies

$$H''(|\Delta z|) = \frac{(c^2 - \sigma^2)^2}{2c\sigma(c^2 + \sigma^2)} \frac{\sinh(\frac{|\Delta z|}{2K_0})}{\cosh^3(\frac{|\Delta w|}{2})} \ge 0. \tag{4.4.28}$$

Proof: Solving equation (4.4.24) for ζ gives

$$\zeta = 2\left(\sinh\left(\frac{|\Delta z|}{2K_0}\right)\right)^2. \tag{4.4.29}$$

Substituting (4.4.29) into (4.4.23) yields (4.4.27), and the formula

$$f_+^{-1}(y) = 2\sinh^2(\ln(y)). \tag{4.4.30}$$

Using this to solve for ζ in (4.4.23), (4.4.24), equating, and taking square roots, gives (4.4.25). Implicitly differentiating (4.4.25) and simplifying gives (4.4.28). □

It follows directly from Lemma 10 that $H(|\Delta z|)$ is a monotone increasing convex up function of $|\Delta z|$ that is superlinear in the sense that

$$|\Delta z| < H(|\Delta z|) < \infty, \tag{4.4.31}$$

for $\Delta z \neq 0$, and

$$\lim_{|\Delta z| \to 0} \frac{H(|\Delta z|)}{|\Delta z|} = 1, \tag{4.4.32}$$

c.f. Figure 4.

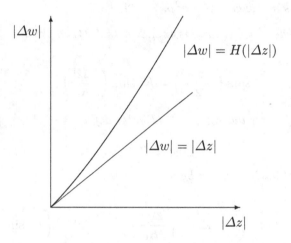

H is increasing, convex up.

Figure 4

Since $|\Delta w| = |\Delta z|$ along all 1- and 2-rarefaction curves, we have the following proposition:

Lemma 11 *Let* $\mathbf{z}_L, \mathbf{z}_R$ *be the left and right states of an elementary wave* γ, *so that*

$$\gamma = [\mathbf{z}_L, \mathbf{z}_R]. \tag{4.4.33}$$

Then

$$|\Delta w| \leq H(|\gamma|), \qquad (4.4.34)$$

where

$$\Delta w = w_R - w_L, \qquad (4.4.35)$$
$$|\gamma| = |\Delta z| = |z_R - z_L|, \qquad (4.4.36)$$

and H is given in (4.4.27).

By the convexity of H and Jenson's inequality we have:

Proposition 11 *Let $\gamma_1, ..., \gamma_n$ be any set of elementary waves. Then*

$$\sum_{i=1}^{n} |\gamma_i| \leq \sum_{i=1}^{n} H(|\gamma_i|) \leq H\left(\sum_{i=1}^{n} |\gamma_i|\right). \qquad (4.4.37)$$

The next Proposition summarizes results in Chapter 2, and follows directly from Proposition 11:

Proposition 12 *For any left and right states z_L, $z_R \in R^2$, there exists a unique solution of the Riemann Problem $[z_L, z_R]$ consisting of a 1-shock or 1-rarefaction wave γ^1, followed by a 2-shock or 2-rarefaction wave γ^2, so that we can write*

$$(\gamma^1, \gamma^2) = [z_L, z_R]. \qquad (4.4.38)$$

The speed of the wave γ^1 is always strictly less than the speed of γ^2, and all wave speeds are subluminous. Moreover, there exist C^2 functions $\Gamma_p : R^2 \to R^2$, one for each $p = 1, 2$, such that $(\gamma^1, \gamma^2) = [z_L, z_R]$ if and only if the vector γ^p satisfies

$$\gamma^p = \Gamma_p(z_R - z_L), \qquad (4.4.39)$$

where,

$$|\gamma^p| = |\Gamma_p(z_R - z_L)| \leq \sqrt{2}\|z_R - z_L\|. \qquad (4.4.40)$$

Proof: The smoothness of Γ_p and the dependence on the difference $z_R - z_L$ follows from the C^2 contact between shock and rarefaction curves, together with the fact that shock wave curves, drawn in the z-plane, are translation invariant. Estimate (4.4.40) can be verified in each of the four cases of the Riemann Problem $[z_L, z_R]$; namely, if $[z_L, z_R]$ is a 1-shock followed by a 2-rarefaction wave *or* a 1-rarefaction wave followed by a 2-shock, then $|\gamma^1| + |\gamma^2| = |z_R - z_L|$. In the other two cases one can verify (4.4.40) assuming that the shock waves lie on the Riemann Invariants, and then see that the divergence of shock and rarefaction curves only improves this estimate. □

We now discuss the boundary Riemann problems posed at mesh points (x_0, t_j), $j = 0, 1, 2, ..$, that lie along the boundary $x_0 = r_0$ in the approximate solution $u_{\Delta x}$. In this case, for fixed j, $u_{0j}^{RP}(x, t)$ is defined in (4.3.8), (4.3.9) as the solution of (4.4.1) together with the initial-boundary data

$$u_0(x) = \left\{ \begin{matrix} v = 0 & x = 0 \\ u_R = u_{0j} & x \geq 0 \end{matrix} \right\}, \qquad (4.4.41)$$

with the origin translated to the bottom center (x_0, t_j) of the mesh rectangle $\mathcal{R}_{0j} \equiv \{(x, t) : x_{r_0} < x \leq x_{\frac{1}{2}}, t_j \leq t < t_{j+1}\}$. Again, vector $\mathbf{A}_{0j}$ is constant on $\mathcal{R}_{0j}$. The following theorem, which generalizes Proposition 9 to include boundary Riemann problems, follows by similar reasoning. (The boundary Riemann problem was not discussed in Chapter 2, but was discussed within the context of a fractional step method in [19]. See also [26, 23].)

Proposition 13 *Assume that u_R lies in the region $\rho > 0$, $-c < v < c$. Then the boundary Riemann problem (4.4.1), (4.4.41) has a unique solution consisting of a single elementary 2-wave γ_{0j}^2 of positive speed. Moreover, the CFL condition (4.3.2) guarantees that the speed of the wave γ_{0j}^2 is always smaller than half the the mesh speed $\frac{\Delta x}{\Delta t} = Max \left\{ 2\sqrt{\frac{A}{B}} \right\}$, and thus γ_{0j}^2 cannot hit the boundary of $\mathcal{R}_{0j}$ within one timestep.*

For fixed $\mathbf{A}_{0j}$, let

$$[0, u_R] \equiv [0, u_R](x, t), \qquad (4.4.42)$$

denote the solution of the Riemann problem (4.4.1), (4.4.41), and write

$$[0, u_R] = \gamma^2, \qquad (4.4.43)$$

to indicate that the solution $[0, u_R](x, t)$ consists of the single wave γ^2, a 2-wave. Again, define the strength of the Riemann problem $[0, u_R]$ as the strength of its elementary wave,

$$|[0, u_R]| = |\gamma^2|. \qquad (4.4.44)$$

The following theorem generalizes Proposition 10 to include the boundary Riemann problems, and this implies that the sum of the strengths of elementary waves are non-increasing during boundary wave interactions, so long as $\mathbf{A}_{ij}$ is constant.

Proposition 14 *Assume that $\mathbf{A}_{ij}$ is fixed. Let u_M, and u_R be any pair of states in the region $\rho > 0$, $-c < v < c$. Then*

$$|[0, u_R]| \leq |[0, u_M]| + |[u_M, u_R]|. \qquad (4.4.45)$$

4.5 The ODE Step

In this section we analyze the ODE step (4.3.22) of the fractional step scheme. Recall that this arises by rewriting system (1.4.3) in the form $u_t + \frac{\partial f}{\partial u} u_x = g - \mathbf{A}' \cdot \nabla_{\mathbf{A}} f \equiv G(\mathbf{A}, u, x)$ and neglecting the flux term containing u_x. Then the jumps in $\mathbf{A}$ at the vertical lines $x_{i+\frac{1}{2}}$, $i = 0, 1, ...$, are accounted for by the $\mathbf{A}' \cdot \nabla_{\mathbf{A}} f$ term on the RHS of this equation. Using (4.3.20), (4.3.21) and the fact that $(u^0, u^1) = (T_M^{00}, T_M^{01})$, system (4.3.22) takes the form

$$\dot{T}_M^{00} = -\frac{1}{2}\sqrt{\frac{A}{B}} T_M^{01} \left\{ \frac{2(B+1)}{x} - \kappa x B \left(T_M^{00} - T_M^{11} \right) \right\} \equiv G^0(\mathbf{A}, u, x), \quad (4.5.1)$$

$$\dot{T}_M^{00} = -\frac{1}{2}\sqrt{\frac{A}{B}} \left\{ \frac{4}{x} T_M^{11} + \frac{(B-1)}{x} \left(T_M^{00} + T_M^{11} \right) \right.$$

$$\left. + \kappa x B \left[T_M^{00} T_M^{11} - 2 \left(T_M^{01} \right)^2 + \left(T_M^{11} \right)^2 \right] - 4x T^{22} \right\} \equiv G^1(\mathbf{A}, u, x). \quad (4.5.2)$$

We now analyze the solution trajectories for system (4.5.1), (4.5.2) in the (ρ, v)-plane. To this end, we record the following identities which are easily derived from (1.3.12),(1.3.13),(1.3.14), and (4.2.6):

$$\left(T_M^{11} \right)^2 - \left(T_M^{01} \right)^2 = \frac{\sigma^4 - v^2 c^2}{c^2 - v^2} \rho^2 c^2, \quad (4.5.3)$$

$$T_M^{00} T_M^{11} - \left(T_M^{01} \right)^2 = \sigma^2 \rho^2 c^2, \quad (4.5.4)$$

$$\left(T_M^{11} \right)^2 - 2 \left(T_M^{01} \right)^2 + T_M^{00} T_M^{11} = \frac{(\sigma^2 - v^2)(c^2 + \sigma^2)}{c^2 - v^2} \rho^2 c^2, \quad (4.5.5)$$

$$T_M^{00} + T_M^{11} = \frac{(c^2 + \sigma^2)(c^2 + v^2)}{c^2 - v^2} \rho, \quad (4.5.6)$$

$$T_M^{00} - T_M^{11} = (c^2 - \sigma^2)\rho. \quad (4.5.7)$$

Using (4.5.3)-(4.5.7) in the RHS of (4.5.1), (4.5.2), we obtain after simplification,

$$G^0 = -\frac{1}{2}\sqrt{\frac{A}{B}} \left(\frac{c^2 + \sigma^2}{c^2 - v^2} \right) cv \frac{\rho}{x} \left\{ 2(B+1) - \kappa B(c^2 - \sigma^2)\rho x^2 \right\}, \quad (4.5.8)$$

$$G^1 = -\frac{1}{2}\sqrt{\frac{A}{B}} \left(\frac{c^2 + \sigma^2}{c^2 - v^2} \right) \frac{\rho}{x} \left\{ 4v^2 + (B-1)(c^2 + v^2) + \kappa B(\sigma^2 - v^2)c^2 \rho x^2 \right\}. \quad (4.5.9)$$

Now differentiating the LHS of (4.5.1), (4.5.2) gives

$$\dot{\rho}\frac{\partial T_M^{00}}{\partial \rho} + \dot{v}\frac{\partial T_M^{00}}{\partial v} = G^0, \tag{4.5.10}$$

$$\dot{\rho}\frac{\partial T_M^{01}}{\partial \rho} + \dot{v}\frac{\partial T_M^{01}}{\partial v} = G^1. \tag{4.5.11}$$

Thus it follows from Cramer's Rule that system (4.5.1), (4.5.2) in (ρ, v)-variables is given by

$$\dot{\rho} = \frac{D_\rho}{D}, \tag{4.5.12}$$

$$\dot{v} = \frac{D_v}{D}, \tag{4.5.13}$$

where

$$D_\rho = \begin{vmatrix} G^0 & \frac{\partial T_M^{00}}{\partial v} \\ G^1 & \frac{\partial T_M^{01}}{\partial v} \end{vmatrix}, \tag{4.5.14}$$

$$D_v = \begin{vmatrix} \frac{\partial T_M^{00}}{\partial \rho} & G^0 \\ \frac{\partial T_M^{01}}{\partial \rho} & G^1 \end{vmatrix}, \tag{4.5.15}$$

$$D = \begin{vmatrix} \frac{\partial T_M^{00}}{\partial \rho} & \frac{\partial T_M^{00}}{\partial v} \\ \frac{\partial T_M^{01}}{\partial \rho} & \frac{\partial T_M^{01}}{\partial v} \end{vmatrix}. \tag{4.5.16}$$

Using (1.3.12) and (1.3.13) we obtain

$$\frac{\partial T_M^{00}}{\partial \rho} = \frac{c^4 + \sigma^2 v^2}{c^2 - v^2},$$

$$\frac{\partial T_M^{00}}{\partial v} = 2\frac{(c^2 + \sigma^2)c^2 v}{(c^2 - v^2)^2}\rho,$$

$$\frac{\partial T_M^{01}}{\partial \rho} = \frac{(\sigma^2 + c^2)cv}{c^2 - v^2},$$

$$\frac{\partial T_M^{01}}{\partial v} = \frac{(\sigma^2 + c^2)(c^2 + v^2)c}{(c^2 - v^2)^2}\rho,$$

and

$$D = \frac{(c^2 + \sigma^2)(c^4 - \sigma^2 v^2)c}{(c^2 - v^2)^2}\rho. \tag{4.5.17}$$

A calculation using these together with (4.5.8), (4.5.9) leads to

$$D_\rho = -\frac{1}{2}\sqrt{\frac{A}{B}}\left(\frac{c^2+\sigma^2}{c^2-v^2}\right)^2\frac{c^2}{x}\{4-\kappa B(c^2+\sigma^2)\rho x^2\}\rho^2,$$

$$D_v = -\frac{1}{2}\sqrt{\frac{A}{B}}\left(\frac{c^2+\sigma^2}{c^2-v^2}\right)\frac{\sigma^2 c^2}{x} \qquad (4.5.18)$$
$$\cdot\left\{-4\frac{v^2}{c^2}+(B-1)\frac{c^4-\sigma^2 v^2}{\sigma^2 c^2}+\kappa B(c^2+v^2)\rho x^2\right\}\rho.$$

Putting (4.5.17)-(4.5.18) into (4.5.12), (4.5.13) and simplfying, we obtain system (4.5.1), (4.5.2) in (ρ, v)-variables:

$$\dot\rho = -\frac{1}{2}\sqrt{\frac{A}{B}}\left(\frac{c^2+\sigma^2}{c^4-\sigma^2 v^2}\right)\frac{vc}{x}\{4-\kappa B(c^2+\sigma^2)\rho x^2\}\rho, \qquad (4.5.19)$$

$$\dot v = -\frac{1}{2}\sqrt{\frac{A}{B}}\left(\frac{c^2-v^2}{c^4-\sigma^2 v^2}\right)\frac{\sigma^2 c}{x}$$
$$\cdot\left\{-4\frac{v^2}{c^2}+(B-1)\frac{c^4-\sigma^2 v^2}{\sigma^2 c^2}+\kappa B(c^2+v^2)\rho x^2\right\},$$

$$(4.5.20)$$

For convenience, we rewrite system (4.5.19), (4.5.20) in the form

$$\dot\rho = \frac{\kappa\sqrt{AB}x}{2}\left[\frac{(c^2+\sigma^2)^2 vc}{c^4-\sigma^2 v^2}\right]\rho\{\rho-\rho_1\}, \qquad (4.5.21)$$

$$\dot v = -\frac{\kappa\sqrt{AB}x}{2}\left[\frac{(c^4-v^4)\sigma^2 c}{c^4-\sigma^2 v^2}\right]\{\rho-\rho_2\}, \qquad (4.5.22)$$

where

$$\rho_1 = \frac{4}{\kappa B(c^2+\sigma^2)x^2}, \qquad (4.5.23)$$

and

$$\rho_2 = \frac{4v^2\sigma^2-(B-1)(c^4-\sigma^2 v^2)}{\kappa B(c^2+v^2)\sigma^2 c^2 x^2}, \qquad (4.5.24)$$

where, (by a simple calculation),

$$\rho_2 < \frac{4v^2\sigma^2}{\kappa B(c^2+v^2)\sigma^2 c^2 x^2} < \rho_1, \qquad (4.5.25)$$

for all values of $v \in (-c, c)$.

We devote the remainder of this section to the proof of the following theorem, which gives a global bound for solutions of $\dot{u} = G(\mathbf{A}, u, x)$, starting from arbitrary initial data

$$u(0) = u_0 \equiv \Psi(\rho_0, v_0), \qquad (4.5.26)$$

assuming that $A > 0$, $B \geq 1$ and $x \geq r_0$ are constant, and assuming the physical bounds $0 < \rho_0 < \infty$, $-c < v_0 < c$, (c.f. (4.2.21)):

Proposition 15 *Assume that A, B and x are constant, that $A > 0$, $B \geq 1$, $x \geq r_0$, and assume that (ρ_0, v_0) satisfies $-c < v_0 < c$ and $0 < \rho_0 < \infty$. Then the solution $(\rho(t), v(t))$ of system (4.5.21), (4.5.22), with initial condition*

$$\rho(0) = \rho_0, \qquad (4.5.27)$$
$$v(0) = v_0, \qquad (4.5.28)$$

exists, is finite, and satisfies

$$-c < v(t) < c,$$

for all $t \geq 0$. Moreover, if $\rho_0 \leq \rho_1$, then $\rho(t) \leq \rho_1$ for all $t \geq 0$; while if $\rho_0 \geq \rho_1$, then

$$\rho_1 \leq \rho(t) \leq \max\left\{ \rho_1, \rho_0 \left(\frac{c^2}{c^2 - v_0^2} \right)^{\frac{1}{2}\left(\frac{c^2 + \sigma^2}{c\sigma} \right)^2} \right\}, \qquad (4.5.29)$$

for all $t > 0$.

Proof: For fixed $\mathbf{A}$ and x, system (4.5.21), (4.5.22) is an autonomous system of the form

$$\dot{\rho} = f_1(\rho, v),$$
$$\dot{v} = f_2(\rho, v).$$

Note that $\rho = 0$ and $v = \pm c$ are solution trajectories for this system. Since the system is autonomous, solution trajectories never intersect, and so it follows that $\rho > 0$, $|v| < c$ is an invariant region for solutions. Note also that since ρ_1 is independent of v, the isocline $\rho \equiv \rho_1$ also defines a solution curve for system (4.5.21), (4.5.22), and so it also cannot be crossed by other solution trajectories. Thus $0 < \rho < \rho_1$, $|v| < c$ is a bounded invariant region, and $\rho > \rho_1$, $|v| < c$ is an unbounded invariant region, for solutions of system (4.5.21),

(4.5.22). Thus it remains only to verify (4.5.29), and it follows that the only obstacle to global existence for the initial value problem (4.5.21), (4.5.22), (4.5.27), (4.5.28), is the case $\rho_0 > \rho_1$, and the possibility that $\rho(t) \to \infty$ before $t \to \infty$. Note that (4.5.21) is quadratic in ρ, so the bound (4.5.29) on ρ is not a consequence of equation (4.5.21) alone. However, (4.5.22) implies that ρ is bounded, as we now show.

If $\rho_0 > \rho_1$, then since $\rho_1 > \rho_2$ for all values of v, it follows that $\dot{v} < 0$ for all time. Consequently, $v(t) \le v_0$, and $\rho(t)$ can only increase while $v \ge 0$. Once v hits $v = 0$, $v(t) < 0$ and $\rho(t)$ decreases from that time forward. Thus it suffices to estimate the change in $\rho(t)$ while $0 \le v(t) \le v_0$. But from (4.5.21), (4.5.22), we have

$$\frac{d\rho}{dv} = -\frac{(c^2 + \sigma^2)^2 v}{(c^4 - v^4)\sigma^2} \frac{\rho - \rho_1}{\rho - \rho_2} \rho \tag{4.5.30}$$

$$\ge -\left(\frac{c^2 - \sigma^2}{\sigma c}\right)^2 \frac{v}{c^2 - v^2}\rho,$$

where we have used $\rho \ge \rho_1 \ge \rho_2$. Integrating this inequality by separation of variables gives the inequality (4.5.29). $\square$

The phase portrait for solutions of (4.5.21), (4.5.22), is given in Figure 4.1.

By the results of Section 4.4 the Riemann problem solutions preserve the bounds $0 < \rho < \infty$, $|v| < c$, and all (invariant) wave speeds remain bounded by c, so long as $0 < \rho < \infty$, $|v| < c$ initially. By the results in this section, it follows that these bounds also are maintained under the ODE step. But (4.3.26) and (4.3.43) imply that that the only way the approximate solution $u_{\Delta x}$ can fail to be defined for all time, is if $B \to \infty$, or the CFL condition fails. The following theorem is a direct consequence of (4.3.26) and (4.3.43):

Proposition 16 *Let $\bar{B}$, $\bar{M}$ denote arbitrary positive constants, let*

$$\Lambda = \frac{\Delta x}{\Delta t} = 2\sqrt{G_{AB}(\bar{B}, \bar{M})},$$

and assume that the initial data $u_0(\cdot)$ satisfies the bounds $0 < \rho < \infty$, $|v| < c$ for all $x \ge r_0$. Then the approximate solution $u_{\Delta x}$ is defined, and continues to satisfy the bounds $0 < \rho < \infty$, $|v| < c$ for all $x \ge r_0$, $t \le t_J$, so long as

$$\|M_{\Delta x}\|_\infty < \bar{M},$$

$$\|B_{\Delta x}\|_\infty < \bar{B},$$

for all $x \ge r_0$, $j \le J$.

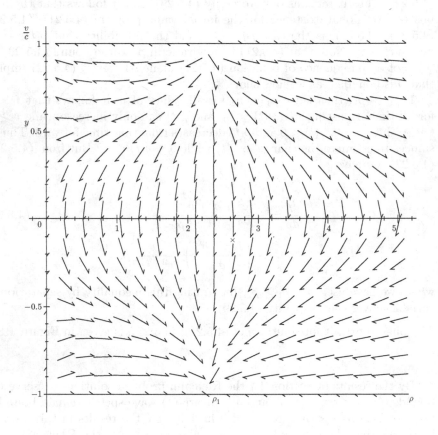

Fig. 4.1. The phase portrait for system (4.5.21), (4.5.22)

As a final comment, note that we have bounds for the RP step, and bounds for the ODE step, but it remains to obtain bounds that apply to both steps. Also, the fact that ρ and v remain finite in each approximate solution does not rule out $\rho \to \infty$ in the actual solution. For this, we need estimates that are independent of Δx, c.f. [19].

4.6 Estimates for the ODE Step

In this section we obtain estimates for the growth of the total variation of $\ln \rho$ and $\ln \frac{c+v}{c-v}$ under the evolution of the ODE $\dot{u} = G(\mathbf{A}, r, x)$, which is equivalent to the system (4.5.21), (4.5.22). To this end, rewrite system (4.5.21), (4.5.22) in terms of the variables $(z, w) \equiv (K_0 \ln \rho, K_0 \ln \frac{c+v}{c-v})$ to obtain, c.f. (4.2.15), (4.2.16),

$$\dot{z} = \frac{4\sqrt{AB}}{x} \left(\frac{\sigma v c^2}{c^4 - \sigma^2 v^2} \right) \left\{ \frac{\kappa(c^2 + \sigma^2)}{4} \rho x^2 - \left(\frac{1}{B} \right)_1 \right\} \equiv F_1(A, B, x, z, w)$$

(4.6.1)

$$\dot{w} = \frac{4\sqrt{AB}}{x} \left(\frac{c^4}{c^4 - \sigma^2 v^2} \right) \left\{ \frac{\kappa(c^2 + v^2)}{4} \frac{\sigma^2}{c^2} \rho x^2 \right.$$
$$\left. - \left(\frac{\sigma^2}{c^2} \left[\frac{1}{B} \frac{v^2}{c^2} - \frac{(B-1)}{4B} \frac{c^4 - \sigma^2 v^2}{\sigma^2 c^2} \right] \right)_2 \right\} \equiv F_2(A, B, x, z, w).$$

(4.6.2)

Here K_0 is defined in (4.2.14), and we use that

$$\dot{z} = K_0 \frac{\dot{\rho}}{\rho},$$
$$\dot{w} = \frac{2c}{c^2 - v^2} \dot{v}.$$

Note also that $\kappa c^2 \rho x^2$ is dimensionless. A calculation shows that the indexed brackets on the RHS of (4.6.1), (4.6.2) satisfy

$$|(\cdot)_i| \le 1, \quad i = 1, 2.$$

(4.6.3)

To verify (4.6.3), use that $B \ge 1$, and

$$|(\cdot)_2| = \frac{\sigma^2}{c^2} \left| \frac{(c^4 + 3\sigma^2 v^2) - B(c^4 - \sigma^2 v^2)}{4B\sigma^2 c^2} \right|$$
$$= \frac{\sigma^2}{c^2} \left| \frac{1}{4} \left(\frac{1}{B} - 1 \right) \left(\frac{c}{\sigma} \right)^2 + \frac{1}{4} \left(\frac{3}{B} + 1 \right) \left(\frac{v}{c} \right)^2 \right|$$
$$\le \frac{\sigma^2}{c^2} Max \left\{ \frac{1}{4} \left(\frac{c}{\sigma} \right)^2, \left(\frac{v}{c} \right)^2 \right\} \le 1.$$

The following theorem gives bounds for the RHS of (4.6.1), (4.6.2).

Proposition 17 *Assume that*

$$1 \le B \le \bar{B},$$

(4.6.4)

$$0 < AB \le G_{AB}(\bar{B}, \bar{M}) \equiv A_{r_0} B_{r_0} \exp \left\{ \frac{8\bar{B}\bar{M}}{r_0} \right\},$$

(4.6.5)

$$S \le \bar{S},$$

(4.6.6)

$$|v| < c,$$

(4.6.7)

and $r_0 \le x < \infty$. Then each of

$$|F_i(A, B, x, z, w)|, \quad \left|\frac{\partial F_i}{\partial z}\right|, \quad \left|\frac{\partial F_i}{\partial w}\right|,$$

$i = 1, 2$, *is bounded by* $\frac{1}{2\sqrt{2}}G_1(\bar{B}, \bar{M}, \bar{S})$, *where* G_1 *is defined by*

$$\frac{1}{2\sqrt{2}}G_1(\bar{B}, \bar{M}, \bar{S}) \equiv \frac{G_0(\kappa c^2 r_0 \bar{S} + 1)}{r_0}, \tag{4.6.8}$$

where

$$G_0 \equiv G_0(\bar{B}, \bar{M}) = K_1\sqrt{G_{AB}(\bar{B}, \bar{M})}, \tag{4.6.9}$$

$$K_1 = \frac{8c^4}{(c^2 - \sigma^2)^2}. \tag{4.6.10}$$

Here we use the notation that K with a subscript denotes a constant that depends only on κ, r_0, σ and c, while $G(\cdot)$ denotes a constant that depends also on $\bar{A}, \bar{B}, \bar{S}$ and $\bar{M}$, whichever appear in the parentheses after the G. We include the factor $(2\sqrt{2})^{-1}$ in (4.6.8) for future convenience, c.f. Theorem 18 and Proposition 19 below.

Proof: This follows by direct calculation, using $|v| < c$, $\sigma < c$. For example,

$$\left|\frac{\partial F_2}{\partial z}\right| = \left|\frac{\partial \rho}{\partial z}\frac{\partial F_2}{\partial \rho}\right| = \left|\frac{\rho}{K_0}\frac{\partial F_2}{\partial \rho}\right|$$

$$= \left|\frac{\rho(c^2 + v^2)}{2\sigma c}\frac{4\sqrt{AB}}{x}\left(\frac{c^4}{c^4 - \sigma^2 v^2}\right)\left\{\frac{\kappa(c^2 + v^2)}{4}\frac{\sigma^2}{c^2}x^2\right\}\right|$$

$$\le \frac{G_0(\bar{B}, \bar{M})}{r_0}(\kappa c^2 r_0 \bar{S} + 1).$$

Also,

$$\left|\frac{\partial F_2}{\partial w}\right| = \left|\frac{\partial v}{\partial w}\frac{\partial F_2}{\partial v}\right| = \left|\left(\frac{c^2 - v^2}{2c}\right)\frac{\partial F_2}{\partial v}\right|$$

$$= \left|\left\{\frac{4\sqrt{AB}}{x}\left(\frac{c^2 - v^2}{2c}\right)\frac{\partial}{\partial v}\left(\frac{c^4}{c^4 - \sigma^2 v^2}\right)\{\cdot\}_*\right\}_3 \right.$$

$$\left. + \left\{\frac{4\sqrt{AB}}{x}\left(\frac{c^3(c^2 - v^2)}{2(c^4 - \sigma^2 v^2)}\right)\frac{\partial}{\partial v}\{\cdot\}_*\right\}_4\right|,$$

where

$$\{\cdot\}_* = \left\{ \frac{\kappa(c^2 + v^2)\,\sigma^2}{4}\frac{\sigma^2}{c^2}\rho x^2 - \left(\frac{\sigma^2}{c^2}\left[\frac{1}{B}\frac{v^2}{c^2} - \frac{(B-1)}{4B}\frac{c^4 - \sigma^2 v^2}{\sigma^2 c^2}\right]\right)_2 \right\}_*.$$

But straightforward estimates show that

$$|\{\cdot\}_i| \le \frac{1}{2}\frac{G_0(\bar{B}, \bar{M})}{r_0}(\kappa c^2 r_0 \bar{S} + 1),$$

for both $i = 1$ and $i = 2$, and so

$$\left|\frac{\partial F_2}{\partial w}\right| \le \frac{G_0(\bar{B}, \bar{M})}{r_0}(\kappa c^2 r_0 \bar{S} + 1).$$

This completes the proof of the theorem.$\Box$

We now study solutions of (4.6.1), (4.6.2) in the $\mathbf{z}$-plane,

$$\mathbf{z} = (z, w) \equiv (K_0 \ln \rho, \ln \frac{c - v}{c + v}), \tag{4.6.11}$$

$$\|\mathbf{z}\| = \sqrt{z^2 + w^2}, \tag{4.6.12}$$

so that system (4.6.1), (4.6.2) can be written as

$$\dot{\mathbf{z}} = F(\mathbf{A}, x, \mathbf{z}), \tag{4.6.13}$$

where $\mathbf{A} = (A, B)$ and

$$F = (F_1, F_2). \tag{4.6.14}$$

Let

$$\mathbf{z}(t) \equiv \mathbf{z}(t; \mathbf{A}, x, \mathbf{z}_0) \tag{4.6.15}$$

denote the solution of (4.6.1), (4.6.2) starting from initial data

$$\mathbf{z}(0) = \mathbf{z}_0, \tag{4.6.16}$$

treating $\mathbf{A}$ and x as constants. We now estimate

$$\frac{d}{dt}\|\mathbf{z}(t)\|. \tag{4.6.17}$$

To start, note first that for any smooth curve $\mathbf{z}(t)$,

$$\left|\frac{d}{dt}\|\mathbf{z}\|\right| = \left|\frac{\mathbf{z}(t) \cdot \dot{\mathbf{z}}(t)}{\|\mathbf{z}(t)\|}\right| \le \|\dot{\mathbf{z}}(t)\|. \tag{4.6.18}$$

Thus, if $\mathbf{z}(t)$ denotes a solution of (4.6.1), (4.6.2), then

$$\left| \frac{d}{dt} \|\mathbf{z}\| \right| \leq \|F(A, B, x, \mathbf{z}(t))\|$$

$$= \sqrt{2} \frac{G_0(\bar{B}, \bar{M})}{r_0} (\kappa c^2 r_0 \bar{S} + 1). \tag{4.6.19}$$

We next obtain a similar estimate for

$$\left| \frac{d}{dt} \|\mathbf{z}_R(t) - \mathbf{z}_L(t)\| \right|, \tag{4.6.20}$$

where

$$\mathbf{z}_L(t) \equiv \mathbf{z}(t; \mathbf{A}, x_L, \mathbf{z}_L), \tag{4.6.21}$$

$$\mathbf{z}_R(t) \equiv \mathbf{z}(t; \mathbf{A}, x_R, \mathbf{z}_R), \tag{4.6.22}$$

and $\mathbf{A}_L, x_L, \mathbf{A}_R, x_R$, are constants. (Here, $\mathbf{z}_L, \mathbf{z}_R$ could be consecutive constant states that pose a Riemann problem in the construction of $u_{\Delta x}$.) Then,

$$\left| \frac{d}{dt} \|\mathbf{z}_R(t) - \mathbf{z}_L(t)\| \right| \leq \|\dot{\mathbf{z}}_R(t) - \dot{\mathbf{z}}_L(t)\|$$

$$= \|F(\mathbf{z}_R, \mathbf{A}, x) - F(\mathbf{z}_L, \mathbf{A}, x)\| = \|\Delta F\|$$

$$\leq \sqrt{2} Max\{|\Delta F_1|, |\Delta F_2|\}. \tag{4.6.23}$$

But

$$|\Delta F_i| \leq \left| \frac{\partial F_i}{\partial z} \right| |\Delta z| + \left| \frac{\partial F_i}{\partial w} \right| |\Delta w|, \tag{4.6.24}$$

so by Proposition 17, if (4.6.4)-(4.6.7) hold, then

$$|\Delta F_i| \leq \frac{G_0(\bar{B}, \bar{M})}{r_0} (\kappa c^2 r_0 \bar{S} + 1) \{|\Delta z| + |\Delta w|\}$$

$$\leq \sqrt{2} \frac{G_0(\bar{B}, \bar{M})}{r_0} (\kappa c^2 r_0 \bar{S} + 1) \{\|\Delta \mathbf{z}\|\}. \tag{4.6.25}$$

We have the following result:

Proposition 18 *Let $\mathbf{z}_L(t), \mathbf{z}_R(t)$ be defined by (4.6.21) and (4.6.22), and assume $x_L < x_R$, and that (4.6.4)-(4.6.7) of Proposition 17 hold. Then*

$$\left| \frac{d}{dt} \|\mathbf{z}_R(t) - \mathbf{z}_L(t)\| \right| \leq \frac{G_1}{\sqrt{2}} \{\|\mathbf{z}_R(t) - \mathbf{z}_L(t)\|\}, \tag{4.6.26}$$

$$\|\mathbf{z}_R(t) - \mathbf{z}_L(t)\| \leq \|\mathbf{z}_R - \mathbf{z}_L\| e^{\frac{G_1}{\sqrt{2}}t}, \qquad (4.6.27)$$

where, c.f. (4.6.8),

$$G_1 \equiv G_1(\bar{B}, \bar{M}, \bar{S}) = 2\sqrt{2}\frac{G_0(\bar{B}, \bar{M})}{r_0}(\kappa c^2 r_0 \bar{S} + 1). \qquad (4.6.28)$$

The states $\mathbf{z}_L(t), \mathbf{z}_R(t)$ pose a Riemann Problem $[\mathbf{z}_L(t), \mathbf{z}_R(t)]$ at each time $t \geq 0$. Let

$$[\mathbf{z}_L(t), \mathbf{z}_R(t)] = (\gamma^1(t), \gamma^2(t)), \qquad (4.6.29)$$

denote the waves that solve this Riemann problem, c.f. (4.4.16).

Lemma 12 *The following estimate holds:*

$$\left|\frac{d}{dt}\sum_{p=1,2}|\gamma^p(t)|\right| \leq \sqrt{2}\|\dot{\mathbf{z}}_R(t) - \dot{\mathbf{z}}_L(t)\|. \qquad (4.6.30)$$

$$\leq G_1\|\Delta \mathbf{z}\|.$$

Proof: For the purposes of the proof, let $\mathbf{z}(t) = \mathbf{z}_R(t) - \mathbf{z}_L(t)$, and let

$$\gamma^p(\mathbf{z}(t)) \equiv \Gamma_p(\mathbf{z}(t)), \qquad (4.6.31)$$

where Γ_p is defined in (4.4.39) of Proposition 12. Then by Propositions 10, 14 and (4.4.40),

$$\sum_{p=1,2}\{|\gamma^p(\mathbf{z}(t))| - |\gamma^p(\mathbf{z}(0))|\} \leq \sum_{p=1,2}|\gamma^p(\mathbf{z}(t) - \mathbf{z}(0))|$$

$$\leq \sqrt{2}\|\mathbf{z}(t) - \mathbf{z}(0)\|. \qquad (4.6.32)$$

Similarly,

$$\sum_{p=1,2}\{|\gamma^p(\mathbf{z}(0))| - |\gamma^p(\mathbf{z}(t))|\} \leq \sum_{p=1,2}|\gamma^p(\mathbf{z}(0) - \mathbf{z}(t))|$$

$$\leq \sqrt{2}\|\mathbf{z}(t) - \mathbf{z}(0)\|. \qquad (4.6.33)$$

Thus (4.6.32) together with (4.6.33) imply that

$$\left|\frac{d}{dt}\sum_{p=1,2}|\gamma^p(\mathbf{z}(t))|\right| \leq \sqrt{2}\|\dot{\mathbf{z}}(t)\|, \qquad (4.6.34)$$

which is (4.6.30). The second inequality in (4.6.30) follows directly from (4.6.26). $\square$

We have the following, c.f. (4.6.8):

Proposition 19 *Let* $(\gamma^1(t), \gamma^2(t)) = [\mathbf{z}_L(t), \mathbf{z}_R(t)]$, *where* $\mathbf{z}_L(t)$, $\mathbf{z}_R(t)$ *solve the ODE (4.6.1), (4.6.2). Assume that (4.6.4)-(4.6.7) of Proposition 17 hold. Then*

$$|\gamma^1(t)| + |\gamma^2(t)| \le |\gamma^1(0)| + |\gamma^2(0)| + \|\mathbf{z}_R(0) - \mathbf{z}_L(0)\| e^{G_1 t} G_1 t, \qquad (4.6.35)$$

where $G_1 \equiv G_1(\bar{B}, \bar{M}, \bar{S})$ *is given in (4.6.28).*

Proof: This follows from (4.6.30) and (4.6.27). □

4.7 Analysis of the Approximate Solutions

Let $u_{\Delta x}(x, t)$ denote an approximate solution generated by the fractional step Glimm method, starting from initial data $u_0(\cdot)$ that satisfies the finite total mass condition

$$M_{\Delta x}(\infty, 0) = M_{r_0} + M_0 \le \infty, \quad M_0 \equiv \frac{\kappa}{2} \int_{r_0}^{\infty} u_{\Delta x}^0(r, 0) r^2 \, dr; \qquad (4.7.1)$$

the condition for initial locally finite total variation,

$$\sum_{\substack{i_1 \le i \le i_2 \\ p = 1,2}} |\gamma_{i,0}^p| < V_0, \qquad (4.7.2)$$

for all i_1, i_2 such that $|x_{i_2} - x_{i_1}| \le L$; the condition that the initial velocity is bounded uniformly away from the speed of light,

$$|v_{\Delta x}(x, 0)| \le \bar{v}_0 < c; \qquad (4.7.3)$$

and the condition that the initial supnorm of $x\rho$ is bounded,

$$S_{\Delta x}(x, 0) \equiv x\rho_{\Delta x}(x, 0) \le \bar{S}_0 < \infty. \qquad (4.7.4)$$

Note that (4.7.1) and (4.7.4) imply that

$$|w_{\Delta x}(x, 0)| \le \left| \ln \left(\frac{c + \bar{v}_0}{c - \bar{v}_0} \right) \right| \equiv \bar{w}_0, \qquad (4.7.5)$$

and

$$|\bar{z}_{\Delta x}(x, 0)| \le \left| K_0 \ln \left(\frac{\bar{S}}{r_0} \right) \right| \equiv \bar{z}_0. \qquad (4.7.6)$$

Assuming (4.7.1)-(4.7.6), our goal is to find estimates for V_j, $\bar{S}_j$ and $T > 0$ such that

$$\sum_{\substack{i_1 \leq i \leq i_2 \\ p = 1, 2}} |\gamma_{ij}^p| < V_j, \tag{4.7.7}$$

and

$$S_j \equiv \sup_{r \geq r_0} x\rho_{\Delta x}(x, t_j) \leq \bar{S}_j \tag{4.7.8}$$

for all $|x_{i_2} - x_{i_1}| \leq L$, $0 \leq t_j \leq T = t_J \leq 1$. Note that (4.7.7) estimates the total variation in z on x-intervals of length L, and (4.7.8) estimates a weighted supnorm. Estimates for the supnorm and local total variation norm of the approximate solution $u_{\Delta x}$, that are uniform in time, are required to apply the Oleinik compactness argument, [19]. Recall that the waves γ_{ij}^p solve the Riemann Problem $[u_{i-1,j}, u_{ij}]$ for system (4.3.8), and that since $\mathbf{A} = \mathbf{A_{ij}}$ on mesh rectangle $\mathcal{R}_{ij}$, it follows that the source $\mathbf{A}$ affects the speeds of the waves γ_{ij}^p, but the states on the waves themselves agree with the solution $[u_{i-1,j}, u_{ij}]$ for the special relativistic Euler equations (4.3.8) when $\mathbf{A} = (1, 1)$.

To start, let Δ_{ij} denote the interaction diamond centered at (x_i, t_j) in the approximate solution $u_{\Delta x}$. In the case $i > 0$, the diamond Δ_{ij} is formed by the points $(x_{i-1} + a_j\Delta x, t_j)$, $(x_i + a_j\Delta x, t_j)$, $(x_i, t_{j-\frac{1}{2}})$, $(x_i, t_{j+\frac{1}{2}})$, and in the case $i = 0$, $\Delta_{0,j}$ is the half-diamond formed at the boundary by the mesh points $(x_0, t_{j+\frac{1}{2}})$, $(x_0, t_{j-\frac{1}{2}})$, $(x_0 + a_j\Delta x, t_j)$, c.f. Figure 4.2. In the case $i > 0$, the waves γ_{ij}^p solve the Riemann Problem $[u_L, u_R]$, where

$$u_L = u_{i-1,j},$$
$$u_R = u_{ij}.$$

We call these the waves that leave the diamond Δ_{ij}, c.f. [10]. The waves that enter the diamond solve the Riemann Problems $[\hat{u}_L, u_{M_1}]$, $[u_{M_1}, u_{M_2}]$, and $[u_{M_2}, \hat{u}_R]$, where

$$u_{M_1} = u_{i-1,j-1},$$
$$u_{M_2} = u_{i,j-1},$$

and

$$\hat{u}_L = u_{i-1,j}^{RP},$$
$$\hat{u}_R = u_{ij}^{RP}. \tag{4.7.9}$$

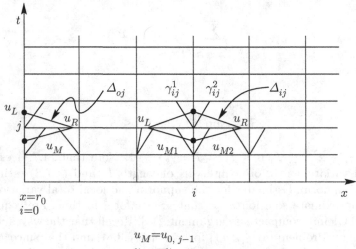

Fig. 4.2. The interaction diamonds Δ_{ij}

States u_L, u_R are obtained from $\hat{u}_L, \hat{u}_R$ by solving the ODE $u_t = G$, written out in (4.3.19). That is, $u_L = \chi^{-1}(\mathbf{z}_L), u_R = \chi^{-1}(\mathbf{z}_R)$, and $\hat{u}_L = \chi^{-1}(\hat{\mathbf{z}}_L), \hat{u}_R = \chi^{-1}(\hat{\mathbf{z}}_R)$, where,

$$\mathbf{z}_L = \mathbf{z}(\Delta t; \mathbf{A}_{ij}, x_i, \hat{\mathbf{z}}_L),$$
$$\mathbf{z}_R = \mathbf{z}(\Delta t; \mathbf{A}_{ij}, x_i, \hat{\mathbf{z}}_R), \tag{4.7.10}$$

c.f. (4.6.15). Thus, using the notation introduced at (4.4.16),

$$[u_L, u_R] = (\gamma_{ij}^1, \gamma_{ij}^2),$$
$$[\hat{u}_L, \hat{u}_R] = (\hat{\gamma}_{ij}^1, \hat{\gamma}_{ij}^2), \tag{4.7.11}$$
$$[u_{M_1}, u_{M_2}] = (\gamma_{i,j-1}^1, \gamma_{i,j-1}^2),$$

and we write

$$[\hat{u}_L, u_{M_1}] = (\gamma_{i-1,j-1}^{R1}, \gamma_{i-1,j-1}^{R2}), \tag{4.7.12}$$
$$[u_{M_2}, u_R] = (\gamma_{i+1,j-1}^{L1}, \gamma_{i+1,j-1}^{L2}).$$

Here we let $\gamma_{ij}^{R1}, \gamma_{ij}^{R2}$ denote the waves in the Riemann Problem posed at (x_i, t_j) that lie to the right of the random point $(x_i + a_{j+1}\Delta x, t_{j+1})$ at time $t = t_{j+1}$, and $\gamma_{ij}^{L1}, \gamma_{ij}^{L2}$ the waves that fall to the left of the random point $(x_{i-1} + a_{j+1}\Delta x, t_{j+1})$, respectively, [26, 10]

In the case of the boundary diamond $\Delta_{0,j}$, the wave γ_{0j}^2 leaves the diamond $\Delta_{0,j}$, and the waves $\gamma_{0,j-1}^2, \gamma_{1,j-1}^{L1}$ and $\gamma_{1,j-1}^{L2}$ enter the diamond. In this case, using the notation introduced at (4.4.44), (4.4.43), we can write

$$[0, u_R] = \gamma_{0j}^2,$$
$$[0, \hat{u}_M] = \hat{\gamma}_{0,j-1}^2,$$
$$[u_M, \hat{u}_R] = (\gamma_{1,j-1}^{L1}, \gamma_{1,j-1}^{L2}), \tag{4.7.13}$$

where

$$u_R = u_{0j},$$
$$\hat{u}_R = u_{0j}^{RP}, \tag{4.7.14}$$
$$u_M = u_{0,j-1}.$$

Now let $|\gamma_{ij}^{IN}|$ denote the sum of the strength of the waves that enter the diamond Δ_{ij}. Thus, if $i > 0$, then

$$|\gamma_{ij}^{IN}| = \sum_{p=1,2} \left\{ |\gamma_{i-1,j-1}^p| + |\gamma_{i,j-1}^p| + |\gamma_{i+1,j-1}^{Lp}| \right\}, \tag{4.7.15}$$

and if $i = 0$,

$$|\gamma_{0j}^{IN}| = |\gamma_{0,j-1}^2| + |\gamma_{1,j}^{L1}| + |\gamma_{1,j}^{L2}|. \tag{4.7.16}$$

It follows from Propositions 10 and 14 that when $i > 0$,

$$|\hat{\gamma}_{ij}^1| + |\hat{\gamma}_{ij}^2| \le |\gamma_{ij}^{IN}|, \tag{4.7.17}$$

and when $i = 0$,

$$|\hat{\gamma}_{0j}^2| \le |\gamma_{0j}^{IN}|. \tag{4.7.18}$$

Now it follows from (4.6.35) of Proposition 19, that if $u_{\Delta x}$ satisfies (4.6.4)-(4.6.7) of Proposition 17, then

$$|\gamma_{ij}^1| + |\gamma_{ij}^2| \le |\hat{\gamma}_{ij}^1| + |\hat{\gamma}_{ij}^2| + \|\hat{z}_R - \hat{z}_L\| e^{G_1 \Delta t} G_1 \Delta t. \tag{4.7.19}$$

But by (4.4.26),

$$|\hat{z}_R - \hat{z}_L| \le |\hat{\gamma}_{ij}^1| + |\hat{\gamma}_{ij}^2| + H(|\hat{\gamma}_{ij}^1|) + H(|\hat{\gamma}_{ij}^2|), \tag{4.7.20}$$

so putting these together we obtain,

$$|\gamma_{ij}^1| + |\gamma_{ij}^2| \le |\hat{\gamma}_{ij}^1| + |\hat{\gamma}_{ij}^2| + \left\{ \sum_{p=1,2} [|\hat{\gamma}_{ij}^p| + H(|\hat{\gamma}_{ij}^p|)] \right\} e^{G_1 \Delta t} G_1 \Delta t. \tag{4.7.21}$$

We can now use (4.7.17), (4.7.18) to estimate $|\hat{\gamma}_{ij}^p|$ and $H(|\hat{\gamma}_{ij}^p|)$.

Note first that since H is convex up, we have Proposition 11, so

$$\sum_{p=1,2} H\left(|\hat{\gamma}_{ij}^p|\right) \leq H\left(\sum_{p=1,2} |\hat{\gamma}_{ij}^p|\right) \leq H(|\gamma_{ij}^{IN}|). \qquad (4.7.22)$$

Let

$$|\gamma_{ij}^{OUT}| = |\gamma_{ij}^1| + |\gamma_{ij}^2|. \qquad (4.7.23)$$

Then putting (4.7.17), (or (4.7.18) at the boundary), and (4.7.22) into (4.7.21), we obtain

$$|\gamma_{ij}^{OUT}| \leq |\gamma_{ij}^{IN}| + \left\{|\gamma_{ij}^{IN}| + H\left(|\gamma_{ij}^{IN}|\right)\right\} e^{G_1 \Delta t} G_1 \Delta t. \qquad (4.7.24)$$

We can also estimate the change in z and $\mathbf{z}$ between (x_i, t_j) and (x_i, t_{j-1}). Since both $\mathbf{z}_{i,j-1}$ and $\mathbf{z}_{ij}^{RP}$ are states on the waves $\gamma_{i,j-1}^p$ or $\gamma_{i+1,j-1}^p$, and by (4.6.19) we know

$$|z_{ij} - z_{ij}^{RP}| \leq \|\mathbf{z}_{ij} - \mathbf{z}_{ij}^{RP}\| \leq G_1 \Delta t, \qquad (4.7.25)$$

it follows that

$$|\mathbf{z}_{ij} - \mathbf{z}_{i,j-1}| \leq \sum_{\substack{l = i, i+1 \\ p = 1, 2}} \left\{|\gamma_{l,j-1}^p| + H\left(|\gamma_{l,j-1}^p|\right)\right\} + G_1 \Delta t, \qquad (4.7.26)$$

and

$$|z_{ij} - z_{i,j-1}| \leq \sum_{\substack{l = i, i+1 \\ p = 1, 2}} |\gamma_{l,j-1}| + G_1 \Delta t. \qquad (4.7.27)$$

We collect our results so far in the following theorem.

Theorem 14 *Let $\bar{M}$, $\bar{B}$, $\bar{S}$, $\bar{v}$, and integer $J_0 > 0$, be any finite positive constants, assume $|\bar{v}| < c$, and let $u_{\Delta x}(x, t)$, $\mathbf{A}_{\Delta x}(x, t)$ be an approximate solution generated by the fractional step Glimm method with*

$$\frac{\Delta x}{\Delta t} = \Lambda = 2\sqrt{G_{AB}(\bar{B}, \bar{M})}. \qquad (4.7.28)$$

Assume that,

$$M_{\Delta x}(x, t_j) \le \bar{M}, \tag{4.7.29}$$
$$B_{\Delta x}(x, t_j) \le \bar{B}, \tag{4.7.30}$$
$$0 < S_{\Delta x}(x, t_j) \le \bar{S} \tag{4.7.31}$$
$$|v_{\Delta x}(x, t_j)| \le \bar{v}, \tag{4.7.32}$$

for all $x \ge r_0$, $0 \le t_j \le T_0 = t_{J_0} \le 1$. Then the speed of each wave γ_{ij}^p generated in $u_{\Delta x}$ at the Riemann Problem step of the method is bounded by the coordinate speed of light $\sqrt{A_{ij}/B_{ij}}$, $i \ge 0$, $0 \le j \le J_0$, and the following estimates hold at each interaction diamond Δ_{ij}, $i \ge 0, j \le J_0 - 1$:

$$\|\mathbf{z}_{ij} - \mathbf{z}_{i,j-1}\| \le \sum_{\substack{l=i,i+1 \\ p=1,2}} \left\{ |\gamma_{l,j-1}^p| + H\left(|\gamma_{l,j-1}^p| \right) \right\} + G_1 \Delta t \tag{4.7.33}$$

$$|z_{ij} - z_{i,j-1}| \le \sum_{\substack{l=i,i+1 \\ p=1,2}} |\gamma_{l,j-1}^p| + G_1 \Delta t, \tag{4.7.34}$$

$$|\gamma_{ij}^{OUT}| - |\gamma_{ij}^{IN}| \le \left\{ |\gamma_{ij}^{IN}| + H\left(|\gamma_{ij}^{IN}| \right) \right\} e^{G_1 \Delta t} G_1 \Delta t, \tag{4.7.35}$$

where,

$$G_1 \equiv G_1(\bar{B}, \bar{M}, \bar{S}) = 2\sqrt{2} \frac{G_0(\bar{B}, \bar{M})}{r_0} (\kappa c^2 r_0 \bar{S} + 1), \tag{4.7.36}$$

$$G_0 \equiv G_0(\bar{B}, \bar{M}) = K_1 \sqrt{G_{AB}(\bar{B}, \bar{M})}, \tag{4.7.37}$$

$$K_1 = \frac{8c^4}{(c^2 - \sigma^2)^2}, \tag{4.7.38}$$

$$G_{AB} = A_{r_0} B_{r_0} \exp\left\{ \frac{8\bar{B}\bar{M}}{r_0} \right\}. \tag{4.7.39}$$

Moreover,

$$1 \le B_{\Delta x}(x, t_j), \tag{4.7.40}$$
$$0 < A_{r_0} < A_{\Delta x}(x, t_j) \le G_{AB}, \tag{4.7.41}$$
$$0 < x u_{\Delta x}^0(x, t_j) \le \frac{c^2 + \sigma^2 \bar{v}^2}{c^2 - \bar{v}^2} \bar{S}, \tag{4.7.42}$$

and

$$|A'_{\Delta x}(x, t_j)| \le \left(\frac{1}{r_0} + \kappa \frac{c^2 + \sigma^2 \bar{v}^2}{c^2 - \bar{v}^2} \bar{S} \right) \bar{B} G_{AB}, \tag{4.7.43}$$

$$|B'_{\Delta x}(x, t_j)| \le \left(\frac{1}{r_0} + \kappa \frac{c^2 + \sigma^2 \bar{v}^2}{c^2 - \bar{v}^2} \bar{S} \right) \bar{B}^2. \tag{4.7.44}$$

for all $x \geq r_0$, $0 \leq t_j \leq T_0 = t_{J_0} \leq 1$.

Note, again, that (4.7.43) gives the Lipschitz continuity in x of $A_{\Delta x}$ and $B_{\Delta x}$ and (4.7.42) implies that $\rho_{\Delta x} > 0$ for $0 \leq t \leq T_0$.

Proof: This follows directly from (4.7.24), (4.7.26) and (4.7.27), together with Propositions 8 and 16.□

Corollary 2 *Assume that the approximate solution $u_{\Delta x}, \mathbf{A}_{\Delta x}$ satisfies the conditions (4.7.28)-(4.7.32) of Theorem 14 up to some time T_0, $0 < T_0 = t_{J_0} \leq 1$, and assume further that there exists constants L, V_0 such that*

$$\sum_{i_1 \leq i \leq i_2,\; p=1,2} |\gamma_{i0}^p| < V_0, \tag{4.7.45}$$

for all $|x_{i_2} - x_{i_1}| \leq L$. Then for any constant $\alpha > 1$:

(A) The following total variation bound holds:

$$\sum_{i_1 \leq i \leq i_2,\; p=1,2} |\gamma_{ij}^p| \leq \alpha \left(1 + \frac{4t_j \sqrt{G_{AB}}}{L} \right) V_0 \leq \alpha \bar{V}_*, \tag{4.7.46}$$

for all $|x_{i_2} - x_{i_1}| \leq L$, so long as $t_j \leq Min\{T_\alpha, T_0\} \leq 1$ where

$$T_\alpha = \left(\frac{1}{G_1 e^{G_1}} \right) \frac{(\alpha - 1)\bar{V}_*}{\{\alpha \bar{V}_* + H(\alpha \bar{V}_*)\}}, \tag{4.7.47}$$

$$\bar{V}_* \equiv \left(1 + \frac{4\sqrt{G_{AB}}}{L} \right) V_0. \tag{4.7.48}$$

(B) The following L_{loc}^1 bounds hold:

$$\int_{x_{i_1}}^{x_{i_2}} \|\mathbf{z}_{\Delta x}(x, t_{j_2}) - \mathbf{z}_{\Delta x}(x, t_{j_1})\| \, dx$$

$$\leq \left\{ 4\sqrt{G_{AB}} \left[\alpha \bar{V}_* + H(\alpha \bar{V}_*) \right] + G_1 |x_{i_2} - x_{i_1}| \right\} |t_{j_2} - t_{j_1}|, \tag{4.7.49}$$

$$\int_{x_{i_1}}^{x_{i_2}} |z_{\Delta x}(x, t_{j_2}) - z_{\Delta x}(x, t_{j_1})| \, dx$$

$$\leq \left\{ 4\sqrt{G_{AB}} \left[\alpha \bar{V}_* \right] + G_1 |x_{i_2} - x_{i_1}| \right\} |t_{j_2} - t_{j_1}|, \tag{4.7.50}$$

for any $t_{j_1} \leq t_{j_2} \leq Min\{T_\alpha, T_0\} \leq 1$, and any $r_0 \leq x_{i_1} \leq x_{i_2} < \infty$.

(C) The following bounds on the supnorm hold:

$$|z_{ij} - z_{i+j,0}| \le \alpha \left(1 + \frac{4t_j\sqrt{G_{AB}}}{L}\right) V_0 + 2\sqrt{G_{AB}}G_1 t_j, \qquad (4.7.51)$$

$$|w_{ij} - w_{i+j,0}| \le H\left(\alpha \left(1 + \frac{4t_j\sqrt{G_{AB}}}{L}\right) V_0\right) + 2\sqrt{G_{AB}}G_1 t_j, \quad (4.7.52)$$

$$\|\mathbf{z}_{ij} - \mathbf{z}_{i+j,0}\| \le \alpha \left(1 + \frac{4t_j\sqrt{G_{AB}}}{L}\right) V_0$$

$$+ H\left(\alpha \left(1 + \frac{4t_j\sqrt{G_{AB}}}{L}\right) V_0\right) + 2\sqrt{G_{AB}}G_1 t_j,$$

$$(4.7.53)$$

for all $t_j \le Min(T_\alpha, T_0) \le 1$.

The motivation for choosing the factor $\left(1 + \frac{4t_j\sqrt{G_{AB}}}{L}\right)$ in (4.7.85) of (A) is that since

$$\frac{\Delta x}{\Delta t} = \Lambda = 2\sqrt{G_{AB}(\bar{B}, \bar{M})}, \qquad (4.7.54)$$

it follows that

$$\left(1 + \frac{4t_j\sqrt{G_{AB}}}{L}\right) \ge \frac{x_{i_2} - x_{i_1} + 4\frac{\Delta x}{\Delta t}t_j}{L}, \qquad (4.7.55)$$

where the RHS of (4.7.55) dominates the number of intervals of length L contained within the domain of dependence of $[x_{i_1}, x_{i_2}]$ at time level t_j. Note that the appearance of $t_j\sqrt{G_{AB}}$ in $\left(1 + \frac{4t_j\sqrt{G_{AB}}}{L}\right)$ (4.7.46) is important because the LHS of (4.7.46) can be estimated independently of $\bar{M}, \bar{B}, \bar{S}$ and $\bar{v}$ for t_j sufficiently small.

Regarding part (C), note that $\mathbf{z}_{i+j,0} = \mathbf{z}_{\Delta x}(x_i + t_j\frac{\Delta x}{\Delta t}, 0)$ where $t_j\frac{\Delta x}{\Delta t} = 2t_j\sqrt{G_{AB}}$ depends on $\bar{M}, \bar{B}$. Note also that since

$$|v_{\Delta x}(x, t_j)| \le \bar{v}_j < c \;\; iff \;\; |w_{\Delta x}(x, t_j)| \le \bar{w}_j = \left|\ln\frac{c + \bar{v}_j}{c - \bar{v}_j}\right|, \qquad (4.7.56)$$

it follows from (4.7.52) that if initially $|w_{\Delta x}(x_i, 0)| \le \bar{w}_0$, then

$$|w_{\Delta x}(x_i, t_j)| \le \bar{w}_0 + H\left(\alpha \left(1 + \frac{4t_j\sqrt{G_{AB}}}{L}\right) V_0\right) + G_1 t_j\sqrt{G_{AB}} = \bar{w}_j,$$

$$(4.7.57)$$

for all $x_i \ge r_0$, $t_j \le Min\{T_\alpha, T_0\} \le 1$. Thus $|w|$ is bounded uniformly and v is bounded uniformly away from c at each $t_j \le Min\{T_\alpha, T_0\} \le 1$ so long as these bounds hold initially.

Proof of (A): Assume $0 < t_j \le T_0$, and consider the interaction diamonds Δ_{ij}, $i = i_1, \ldots, i_2$. Then by (4.7.35), (if $i_1 > 0$),

$$\sum_{i_1 \leq i \leq i_2,\, p=1,2} |\gamma_{ij}^p| = \sum_{i_1 \leq i \leq i_2} |\gamma_{ij}^{OUT}| \tag{4.7.58}$$

$$\leq \sum_{i_1 \leq i \leq i_2} |\gamma_{ij}^{IN}| + \sum_{i_1 \leq i \leq i_2} \left\{ |\gamma_{ij}^{IN}| + H\left(|\gamma_{ij}^{IN}|\right) \right\} e^{G_1 \Delta t} G_1 \Delta t$$

$$\leq \sum_{i_1-1 \leq i \leq i_2+1} |\gamma_{i,j-1}| + \sum_{i_1-1 \leq i \leq i_2+1} \left\{ |\gamma_{i,j-1}| + H\left(|\gamma_{i,j-1}|\right) \right\} e^{G_1 \Delta t} G_1 \Delta t.$$

More generally, let

$$V_j = \sum_{i_1 \leq i \leq i_2} |\gamma_{ij}^{OUT}|, \tag{4.7.59}$$

$$V_{j-1} = \sum_{\partial(i_1-1) \leq i \leq i_2+1} |\gamma_{i,j-1}^{OUT}|, \tag{4.7.60}$$

$$V_0 = \sum_{\partial(i_1-j) \leq i \leq i_2+j} |\gamma_{i,0}^{OUT}|, \tag{4.7.61}$$

where to account for the boundary at $r = r_0$, we let

$$\partial(i_1 - j) = \begin{cases} 0 & i_1 - j \leq 0 \\ i_1 - j & otherwise \end{cases}. \tag{4.7.62}$$

Then (4.7.58) together with the convexity of H imply that

$$V_k - V_{k-1} \leq \left\{ V_{k-1} + H\left(V_{k-1}\right) \right\} e^{G_1 \Delta t} G_1 \Delta t, \tag{4.7.63}$$

for all $k \leq j$. To estimate V_j, define

$$\bar{V}_0 = \left\{ 1 + \frac{x_{i_2+j} - x_{\partial(i_1-j)}}{L} \right\} V_0 \geq \sum_{\partial(i_1-j) \leq i \leq i_2+j,\, p=1,2} |\gamma_{i0}^p|, \tag{4.7.64}$$

c.f. (4.7.45), and inductively let

$$\bar{V}_k = \bar{V}_{k-1} + \left\{ V_{k-1} + H\left(V_{k-1}\right) \right\} e^{G_1 \Delta t} G_1 \Delta t, \tag{4.7.65}$$

to define $\bar{V}_k$ for all $k \leq j$. Note that

$$\bar{V}_* \equiv \left(1 + \frac{4\sqrt{G_{AB}}}{L} \right) V_0 \geq \left(1 + \frac{4 t_j \sqrt{G_{AB}}}{L} \right) V_0 \geq \bar{V}_0, \tag{4.7.66}$$

where we use that

$$\left\{1 + \frac{|x_{i_2+j} - x_{\partial(i_1-j)}|}{L}\right\} \le \left\{\frac{L + 2t_j\frac{\Delta x}{\Delta t}}{L}\right\} \qquad (4.7.67)$$

dominates the number of intervals of length L contained in $|x_{i_2+j} - x_{\partial(i_1-j)}|$. Now $\bar{V}_k$ increases with k, and by induction using (4.7.63) it follows that

$$\bar{V}_k \ge V_k, \qquad (4.7.68)$$

for all $k \le j$. Thus to estimate V_j, it suffices to estimate $\bar{V}_j$. To this end, fix $\alpha > 1$, and let T_α be given by (4.7.47).

Claim: $\bar{V}_j \le \alpha\bar{V}_0$ for all $t_j \le Min\{T_\alpha, T_0\} \le 1$.

To prove the claim, assume that $t_j \le Min\{T_\alpha, T_0\} \le 1$, and t_{j+1} is the first time such that

$$\bar{V}_{j+1} > \alpha\bar{V}_0. \qquad (4.7.69)$$

Then for $t_k \le t_j$,

$$\bar{V}_k - \bar{V}_{k-1} \le \{\alpha\bar{V}_0 + H\left(\alpha\bar{V}_0\right)\} e^{G_1\Delta t}G_1\Delta t, \qquad (4.7.70)$$

and summing we obtain

$$\bar{V}_k - \bar{V}_0 \le \{\alpha\bar{V}_0 + H\left(\alpha\bar{V}_0\right)\} e^{G_1\Delta t}G_1 t_j. \qquad (4.7.71)$$

But solving for t_j in (4.7.71) shows that

$$\{\alpha\bar{V}_0 + H\left(\alpha\bar{V}_0\right)\} e^{G_1\Delta t}G_1 t_j \le (\alpha - 1)\bar{V}_0, \qquad (4.7.72)$$

so long as

$$t_j \le \left(\frac{1}{G_1 e^{G_1}}\right)\frac{(\alpha - 1)\bar{V}_0}{\{\alpha\bar{V}_0 + H(\alpha\bar{V}_0)\}}.$$

But

$$T_\alpha \le \left(\frac{1}{G_1 e^{G_1}}\right)\frac{(\alpha - 1)\bar{V}_0}{\{\alpha\bar{V}_0 + H(\alpha\bar{V}_0)\}}, \qquad (4.7.73)$$

and so it follows (inductively) from (4.7.73) that the bound (4.7.69) is maintained so long as $t_j \le Min\{T_\alpha, T_0\} \le 1$, as claimed.

In light of (4.7.66), it follows that

$$\sum_{i_1 \le i \le i_2} |\gamma_{ij}^p| = V_j \le \bar{V}_j \le \alpha\bar{V}_0 < \alpha\left(1 + \frac{4t_j\sqrt{G_{AB}}}{L}\right)V_0 \le \alpha\bar{V}_*$$

for all $t_j \leq Min\{T_\alpha, T_0\} \leq 1$, which is (4.7.46). The proof of (A) is complete.

Proof of (B): For (4.7.49), estimate as follows:

$$
\int_{x_{i_1}}^{x_{i_2}} \|\mathbf{z}_{\Delta x}(x, t_{j_2}) - \mathbf{z}_{\Delta x}(x, t_{j_1})\| \, dx
$$

$$
= \sum_{i=i_1}^{i_2-1} \|\mathbf{z}_{ij_2} - \mathbf{z}_{ij_1}\| \, \Delta x \leq \sum_{i=i_1}^{i_2-1} \sum_{j=j_1+1}^{j_2} \|\mathbf{z}_{ij} - \mathbf{z}_{i,j-1}\| \Delta x
$$

$$
\leq \sum_{i=i_1}^{i_2-1} \sum_{j=j_1+1}^{j_2} \left\{ \sum_{\substack{l = i, i+1 \\ p = 1, 2}} \left[|\gamma_{l,j-1}^p| + H\left(|\gamma_{l,j-1}^p|\right) \right] + G_1 \Delta t \right\} \Delta x
$$

$$
\leq 2 \sum_{j=j_1+1}^{j_2} \left[\sum_{i=i_1}^{i_2} \sum_{p=1,2} \{|\gamma_{i,j-1}^p| + H\left(|\gamma_{i,j-1}^p|\right)\} \right] \Delta x + G_1 |x_{i_2} - x_{i_1}||t_{j_2} - t_{j_2}|
$$

$$
\leq \left\{ 2 \left[\alpha \bar{V}_* + H(\alpha \bar{V}_*) \right] \frac{\Delta x}{\Delta t} + G_1 |x_{i_2} - x_{i_1}| \right\} |t_{j_2} - t_{j_2}|
$$

where we have used (4.7.33) and (4.7.46). In light of (4.7.28), this verifies (4.7.49). Inequality (4.7.50) follows by the same argument using (4.7.34) in place of (4.7.33).

Proof of (C): Note first that (4.7.33) and (4.7.34) directly imply that

$$
\|\mathbf{z}_{ij} - \mathbf{z}_{i0}\| \leq \sum_{\substack{l = i, i+1 \\ 0 \leq k \leq j-1 \\ p = 1, 2}} \{|\gamma_{lk}^p| + H\left(|\gamma_{lk}^p|\right)\} + G_1 t_j, \qquad (4.7.74)
$$

$$
|z_{ij} - z_{i0}| \leq \sum_{\substack{l = i, i+1 \\ 0 \leq k \leq j-1 \\ p = 1, 2}} |\gamma_{lk}^p| + G_1 t_j. \qquad (4.7.75)
$$

Unfortunately, we cannot use (4.7.74) directly to estimate $\|\mathbf{z}_{ij} - \mathbf{z}_{i0}\|$ because we cannot bound the right-hand-side by V_0 without introducing wave-tracing to identify waves at time t_j with waves at time $t = 0$. To get around this, we estimate $\|\mathbf{z}_{ij} - \mathbf{z}_{i0}\|$ as follows.

Let (x_i, t_j) be fixed. Let J_{ij}^R denote the piecewise linear I-curve that connects mesh points $(x_i, t_{j+\frac{1}{2}})$ to $(x_i + a_j \Delta x_j, t_j)$ to $(x_{i+1}, t_{(j-1)+\frac{1}{2}})$ and so on, continuing downward and to the right until you reach $(x_{i+j} + a_0 \Delta x, 0)$ at time $t_0 = 0$. Let J_{ij}^L connect $(x_i, t_{j+\frac{1}{2}})$ to $(x_{i-1} + a_j \Delta x, t_j)$ to $(x_{i-1}, t_{(j-1)+\frac{1}{2}})$ and so on, continuing downward and to the left until one reaches $t = 0$ at (x_{i-j-1}, t_0) or else stop at $r = r_0$ at the point $(r_0, t_{j_0+\frac{1}{2}})$, (see Figure 4.3).

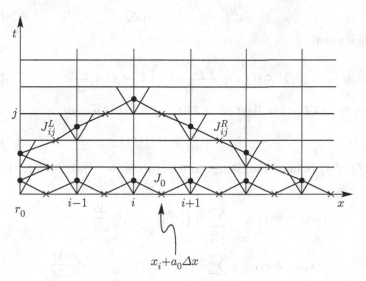

Fig. 4.3. The I-curves J_0, J_{ij}^L and J_{ij}^R

Let J_{ij} denote the I-curve $J_{ij} = J_{ij}^L \bigcup J_{ij}^R$, and recall from [10, 26, 27], that one can connect J_{ij} by a sequence of I-curves, $J_0, \ldots, J_N = J_{ij}$ such that J_{k+1} is an immediate successor of J_k, and J_0 is the I-curve that crosses the waves γ_{i0}^p between $i = \partial(i - i_1)$ and $i = i + j$. (Again, see Figure 4.3.) Since J_k differs from J_{k+1} by a single interaction diamond, it follows by induction using (4.7.58), and the argument (4.7.58)-(4.7.73), that

$$\sum_{J_{ij}} |\gamma_{ij}^p| \le \alpha \left(1 + \frac{4t_j \sqrt{G_{AB}}}{L} \right) V_0, \qquad (4.7.76)$$

where $\sum_{J_{ij}} |\gamma_{ij}^p|$ denotes the sum of the waves γ_{ij}^p that cross the curve J_{ij}. (We have used the assumption $t_j \le Min\{T_\alpha, T_0\} \le 1$.) From this it follows that

$$\sum_{J_{ij}^R} |\gamma_{ij}^p| \le \alpha \left(1 + \frac{4t_j \sqrt{G_{AB}}}{L} \right) V_0. \qquad (4.7.77)$$

But $\sum_{J_{ij}^R} |\gamma_{ij}^p|$ bounds the total variation in z between the state $\mathbf{z}_{i+j,0}$ and the state $\mathbf{z}_{ij}$, except for the change in z that occurs between $\mathbf{z}_{i'j'}^{RP}$ and $\mathbf{z}_{i'j'}$ at each $(x_{i'}, t_{j'})$ that lies on the I-curve J_{ij}^R. But by (4.7.25), we know that

$$\|z_{i'j'}^{RP} - z_{i'j'}\| \le G_1 \Delta x, \tag{4.7.78}$$

so it follows that

$$|z_{ij} - z_{i+j,0}| \le \alpha \left(1 + \frac{4t_j \sqrt{G_{AB}}}{L}\right) V_0 + G_1 t_j \frac{\Delta x}{\Delta t}. \tag{4.7.79}$$

which verifies (4.7.51) in light of (4.7.54), (4.7.54). Also, since

$$\|\gamma_{ij}^p\| \le |\gamma_{ij}^p| + H\left(|\gamma_{ij}^p|\right), \tag{4.7.80}$$

where $H\left(|\gamma_{ij}^p|\right)$ bounds the change in w across wave γ_{ij}^p, it follows that

$$|w_{ij} - w_{i+j,0}| \le \sum_{J_{ij}^R} H\left(|\gamma_{ij}^p|\right) + G_1 t_j \frac{\Delta x}{\Delta t},$$

$$\|\mathbf{z}_{ij} - \mathbf{z}_{i+j,0}\| \le \sum_{J_{ij}^R} |\gamma_{ij}^p| + H\left(|\gamma_{ij}^p|\right) + G_1 t_j \frac{\Delta x}{\Delta t},$$

and so using (4.7.77), (which again uses $t_j \le Min\{T_\alpha, T_0\} \le 1$), we obtain (4.7.52) and (4.7.53). This completes the proof of Corollary 2. $\square$

In order to unify the estimates in (B) and (C), assume that $|x_{i_2} - x_{i_1}| \le L$, and set $G_2 \equiv G_2(\bar{B}, \bar{M}, \bar{S})$ equal to

$$G_2 = Max\left\{4\sqrt{G_{AB}}\left[\alpha\bar{V}_* + H(\alpha\bar{V}_*)\right] + G_1 L, 2\sqrt{G_{AB}}G_1, G_1 e^{G_1}, \sqrt{G_{AB}}\right\}, \tag{4.7.81}$$

where $G_1 \equiv G_1(\bar{B}, \bar{M}, \bar{S})$, $G_{AB} \equiv G_{AB}(\bar{B}, \bar{M})$ and $\bar{V}_*$ are defined in (4.7.36), (4.7.39) and (4.7.48), respectively. (Note that $\bar{V}_*$, and hence G_2, also depend on V_0, but for our purposes we only keep track of the dependence on $\bar{M}, \bar{B}, \bar{S}, \bar{v}$, the constants that are not yet determined by the initial data.) Then the following corollary is a simplification of Corollary 2.

Corollary 3 *Assume that the approximate solution $u_{\Delta x}, \mathbf{A}_{\Delta x}$ satisfies the conditions (4.7.28)-(4.7.32) of Theorem 14 up to some time T_0, $0 < T_0 = t_{J_0} \le 1$, and assume that there exists constants L, V_0 such that*

$$\sum_{i_1 \le i \le i_2,\, p=1,2} |\gamma_{i0}^p| < V_0, \tag{4.7.82}$$

for all $|x_{i_2} - x_{i_1}| \le L$, and assume that $\alpha = 2$, c.f. (4.7.46). Then:

(A) The following total variation bound holds:

$$\sum_{i_1 \leq i \leq i_2, \, p=1,2} |\gamma_{ij}^p| < 2\bar{V}_*, \tag{4.7.83}$$

for all $|x_{i_2} - x_{i_1}| \leq L$, so long as $t_j \leq Min\{T_2, T_0\} \leq 1$, where

$$T_2 = \left(\frac{1}{G_2}\right) \frac{\bar{V}_*}{\{2\bar{V}_* + H(2\bar{V}_*)\}}, \tag{4.7.84}$$

$$\bar{V}_* = \left(1 + \frac{4\sqrt{G_{AB}}}{L}\right) V_0. \tag{4.7.85}$$

(B) The following L_{loc}^1 bounds hold:

$$\int_{x_{i_1}}^{x_{i_2}} \|\mathbf{z}_{\Delta x}(x, t_{j_2}) - \mathbf{z}_{\Delta x}(x, t_{j_1})\| \, dx \leq G_2 |t_{j_2} - t_{j_1}|, \tag{4.7.86}$$

and

$$\int_{x_{i_1}}^{x_{i_2}} |z_{\Delta x}(x, t_{j_2}) - z_{\Delta x}(x, t_{j_1})| \, dx \leq G_2 |t_{j_2} - t_{j_1}|, \tag{4.7.87}$$

for all $r_0 \leq x_{i_1} < x_{i_2} < \infty$, $|x_{i_2} - x_{i_1}| \leq L$, $t_j \leq Min\{T_0, T_2\}$.

(C) The following bounds on the supnorm hold.

$$|z_{ij} - z_{i+j,0}| \leq F_0^*(G_2 \cdot t_j), \tag{4.7.88}$$

$$|w_{ij} - w_{i+j,0}| \leq F_0^*(G_2 \cdot t_j), \tag{4.7.89}$$

$$\|\mathbf{z}_{ij} - \mathbf{z}_{i+j,0}\| \leq F_0^*(G_2 \cdot t_j), \tag{4.7.90}$$

for all $x_i \geq r_0$, $t_j \leq Min\{T_0, T_2\}$, where

$$F_0^*(\xi) = 2\left(1 + \frac{4\xi}{L}\right) V_0 + H\left(2\left(1 + \frac{4\xi}{L}\right) V_0\right) + \xi. \tag{4.7.91}$$

Again, consistent with our notation, the functions $G_2(\cdot, \cdot, \cdot)$ and $F_0^*(\xi)$ depend only on constants σ, c, K_0, r_0, L and V_0 that depend only on the initial data, and so the functions $G_2(\cdot, \cdot, \cdot)$ and $F_0^*(\xi)$ are independent of the constants $\bar{M}, \bar{B}, \bar{S}, \bar{v}$. The functions $G_2(\cdot, \cdot, \cdot)$ and $F_0^*(\xi)$ are also increasing functions of each argument. The main point is that constants that depend on $\bar{M}, \bar{B}, \bar{S}$ or $\bar{v}$ in the estimates (4.7.86)-(4.7.90), are organized into the single constant G_2, (which happens to be independent of $\bar{v}$), and which is always multiplied by the factor t_j. Thus estimates independent of $\bar{M}, \bar{B}, \bar{S}$ and $\bar{v}$ can obtained by making t_j sufficiently small. Note that the formula for $F_0^*(\xi)$ is obtained by substituting 2 for α, and ξ for $t_j\sqrt{G_{AB}}$ and $2\sqrt{G_{AB}}G_1 t_j$, on the RHS of (4.7.53).

4.8 The Elimination of Assumptions

In this section we show that the assumptions (4.7.29)-(4.7.32) in Corollary 3, Theorem 14 above, needn't be assumed, but are implied by values of $\bar{M}, \bar{B}, \bar{S}, \bar{v}$ that can be defined in terms of the initial data alone, subject to restrictions on the time T_0. Once we succeed with this replacement, Theorem 14 and Corollary 3 provide the uniform bounds required to apply the Oleinik compactness argument demonstrating the compactness of approximate solutons up to some finite time T. To start, consider first the bound (4.7.32) for v. Since the constant G_1 in Corollary 3 is independent v, it follows that we can achieve (4.7.32) for a value of $\bar{v}$ defined in terms of the bound on v at time $t = 0$. Indeed, assume that the initial data $v_{\Delta x}(x, 0)$ satisfies

$$|v_{\Delta x}(x,0)| \leq \bar{v}_0 < c \iff |w_{\Delta x}(x,0)| \leq \left| \ln \left(\frac{c + \bar{v}_0}{c - \bar{v}_0} \right) \right| \equiv \bar{w}_0, \quad (4.8.1)$$

for all $r_0 \leq x$. Then assuming the hypotheses of Corollary 3, it follows from (4.7.89) that

$$|w_{\Delta x}(x, t_j)| \leq \bar{w}_0 + F_0^*(G_2 t_j), \quad (4.8.2)$$

for all $r_0 \leq x$, $t_j \leq Min\{T_2, T_0\}$. Therefore, if we *define* $\bar{v}$ so that $\bar{w} = \left| \ln \left(\frac{c+\bar{v}}{c-\bar{v}} \right) \right|$, where

$$\bar{w} \equiv \bar{w}_0 + F_0^*(G_2 t_j), \quad (4.8.3)$$

then (4.7.32) is a *consequence* of our other assumptions. Indeed, to make a rigorous proof out of this, just define $\bar{v}$ by (4.8.3), (4.8.2), and let T_v be the first time at which $|v| \leq \bar{v}$ fails. The argument that leads to the choice of $\bar{w}$ in (4.8.2) then shows that $T_v \geq Min\{T_2, T_0\}$.

Similarly, we now use (4.7.88) to show that $\bar{S}$ can be defined in terms of an initial bound $\bar{S}_0$ on $S_{\Delta x}(x, 0)$ in such a way that (4.7.31) can be eliminated as an assumption in Corollary 3 because it follows as a consequence of our other assumptions. In this case, however, (as in the case of $\bar{M}$ and $\bar{B}$), the constant G_1 depends on $\bar{S}$, so we need a corresponding restriction $t \leq T_S$ for some $T_S \ll 1$. Indeed, assume that the initial data $S_{\Delta x}(x, 0)$ satisfies

$$0 < S_{\Delta x}(x,0) \leq \bar{S}_0, \quad (4.8.4)$$

for all $r_0 \leq x$. Then assuming the hypotheses of Corollary 3, it follows from (4.7.88) that

$$K_0 \ln \rho_{ij} - K_0 \ln \rho_{i+j,0} \leq F_0^*(G_2 \cdot t_j), \quad (4.8.5)$$

and so

$$0 < \rho_{ij} \le F_1^* \left(G_2 \cdot t_j \right) \rho_{i+j,0} \tag{4.8.6}$$

where

$$F_1^*(\xi) = \exp\left\{ \frac{F_0(\xi)}{K_0} \right\} \ge 1. \tag{4.8.7}$$

It follows from (4.8.4) that

$$S_{ij} = x_i \rho_{ij} \le F_1^* \left(G_2 \cdot t_j \right) x_i \rho_{i+j,0} \le F_1^* \left(G_2 \cdot t_j \right) \bar{S}_0. \tag{4.8.8}$$

Inequality (4.8.8) tells us that if we choose $\bar{S} \ge F_1^*(0) \bar{S}_0$, say choose

$$\bar{S} = 2F_1^*(0) \bar{S}_0, \tag{4.8.9}$$

and set

$$T_S = Sup\{t : F_1^* \left(G_2 \cdot t_j \right) \le 2F_1^*(0), \ \text{for all } t_j \le t\}, \tag{4.8.10}$$

then assumption (4.7.29) of Corollary 3, Theorem 14, (that $0 < S_{\Delta x}(x,t) \le \bar{S}$), can be replaced by the condition that $\bar{S}$ is defined in (4.8.9), together with the assumption that $t_j \le T_S$, where T_S is defined in (4.8.9). Note that this argument relies on the fact that the function $F_1^*(\cdot)$ is independent of $\bar{S}$.

We now derive formulas analogous to (4.8.9), (4.8.10), for $\bar{M}, T_M$ and $\bar{B}, T_B$, so that assumptions (4.7.29) and (4.7.30) of Corollary 3, Theorem 14, can be replaced by the condition that $\bar{M}, \bar{B}$ be defined by the values given in the formulas, together with $t_j \le T_M$ and $t_j \le T_B$, respectively. So consider next the total mass

$$M_{\Delta x}(\infty, t_j) = M_{r_0} + M_j, \quad M_j \equiv \frac{\kappa}{2} \int_{r_0}^{\infty} u_{\Delta x}^0(r, t_j) r^2 \, dr. \tag{4.8.11}$$

Using $u^0 = T_M^{00}$ and $\ln w = \frac{c+v}{c-v}$ in (1.3.12), we obtain

$$u^0 = \frac{1}{2} \left\{ (1 + \sigma^2) \cosh w + (1 - \sigma) \right\} \rho. \tag{4.8.12}$$

Thus it follows from (4.8.3) and (4.8.6) that

$$u_{\Delta x}^0(x_i, t_j) \le \frac{1}{2} \left\{ (1 + \sigma^2) \cosh \bar{w} + (1 - \sigma) \right\} F_1^*(G_2 \cdot t_j) \rho_{\Delta x}(x_{i+j}, 0). \tag{4.8.13}$$

Using this in (4.8.11) we obtain

$$M_j \le M_{r_0} + F_2^* \left(G_2 \cdot t_j \right) \frac{\kappa}{2} \int_{r_0}^{\infty} \rho_{\Delta x}(r + j\Delta x, 0) r^2 \, dr$$

$$\le M_{r_0} + F_2^* \left(G_2 \cdot t_j \right) \frac{\kappa}{2} \int_{r_0}^{\infty} u_{\Delta x}^0(r, 0) r^2 \, dr$$

$$= M_{r_0} + F_2^* \left(G_2 \cdot t_j \right) M_0, \tag{4.8.14}$$

where

$$F_2^*(\xi) \equiv \frac{1}{2}\left\{(1+\sigma^2)\cosh(\bar{w}) + (1-\sigma)\right\} F_1^*(\xi), \tag{4.8.15}$$

$$M_0 = \frac{\kappa}{2}\int_{r_0}^{\infty} u_{\Delta x}^0(r,0)r^2\,dr. \tag{4.8.16}$$

Inequality (4.8.14) tells us that if we choose $\bar{M} \geq M_{r_0} + F_2^*(0)M_0$, say choose

$$\bar{M} = M_{r_0} + 2F_2^*(0)M_0, \tag{4.8.17}$$

and set

$$T_M = Sup\{t : M_{r_0} + F_2^*(G_2 \cdot t_j)M_0 \leq \bar{M}, \text{ for all } t_j \leq t\}. \tag{4.8.18}$$

then assumption (4.7.29) of Corollary 3, Theorem 14, can be replaced by the condition that $\bar{M}$ is defined in (4.8.17), together with the assumption that $t_j \leq T_M$, where T_M is defined in (4.8.18).

We now turn to the problem of defining $\bar{B}, T_B$ so as to replace the final assumption (4.7.30) of Theorem 13. Since

$$B_{\Delta x}(x,t) = \frac{1}{1 - \frac{2GM_{\Delta x}(x,t)}{x}}, \tag{4.8.19}$$

it follows that to accomplish this, we must estimate the change in $M_{\Delta x}(x,t_j)$ between times $t = 0$ and $t = t_j$, assuming that Corollary 3, Theorem 14, applies. More generally, assume that (4.7.29)-(4.7.32), and hence Theorem 14, hold up to time t_j, and assume that $0 \leq t_{j_0} < t_j$. We estimate

$$|M_{\Delta x}(x,t_j) - M_{\Delta x}(x,t_{j_0})| \leq \frac{\kappa}{2}\int_{r_0}^{x} |u_{\Delta x}^0(r,t_j) - u_{\Delta x}^0(r,t_{j_0})|r^2\,dr. \tag{4.8.20}$$

To start, let

$$\begin{aligned}
\Delta u^0 &= u_{\Delta x}^0(r,t_j) - u_{\Delta x}^0(r,t_{j_0}), \\
\Delta w &= w_{\Delta x}(r,t_j) - w_{\Delta x}(r,t_{j_0}), \\
\Delta \rho &= \rho_{\Delta x}^0(r,t_j) - \rho_{\Delta x}^0(r,t_{j_0}),
\end{aligned} \tag{4.8.21}$$

etc. Then

$$|\Delta u^0| \leq \left\|\frac{\partial u^0}{\partial w}\right\|_{\infty}|\Delta w| + \left\|\frac{\partial u^0}{\partial z}\right\|_{\infty}|\Delta z|. \tag{4.8.22}$$

From (4.8.12) we calculate

$$\frac{\partial u^0}{\partial z} = \frac{1}{2K_0} \left\{ (1 + \sigma^2) \cosh w + (1 - \sigma) \right\} \rho, \tag{4.8.23}$$

$$\frac{\partial u^0}{\partial w} = \frac{1 + \sigma^2}{2} (\sinh w) \rho. \tag{4.8.24}$$

Since $\left| \frac{\partial u^0}{\partial w} \right| \leq \frac{\partial u^0}{\partial z}$, it follows from (4.8.22) that

$$|\Delta u^0| \leq \frac{1}{K_0} \frac{1}{2} \left\{ (1 + \sigma^2) \cosh(\bar{w}) + (1 - \sigma) \right\} \|\rho\|_\infty \left\{ |\Delta w| + |\Delta z| \right\}$$

$$\leq \frac{\sqrt{2}}{K_0} \frac{F_2^* (G_2 \cdot t_j)}{F_1^* (G_2 \cdot t_j)} \|\rho\|_\infty \|\Delta \mathbf{z}\|. \tag{4.8.25}$$

Putting (4.8.25) into (4.8.20) and using

$$\rho_{\Delta x}(x, t) \leq F_1^* (G_2 \cdot t_j) \rho_{\Delta x} (x + j\Delta x, 0), \tag{4.8.26}$$

we obtain

$$|M_{\Delta x}(x, t_j) - M_{\Delta x}(x, t_{j_0})| \leq \frac{\kappa F_2^* (G_2 \cdot t_j)}{\sqrt{2} K_0} \int_{r_0}^{x} \rho_{\Delta x} (r + j\Delta x, 0) \|\Delta \mathbf{z}\| r^2 \, dr. \tag{4.8.27}$$

We use (4.8.27) again below, but for now we can continue from (4.8.27) to obtain

$$\frac{\kappa F_2^*(G_2 \cdot t_j)}{\sqrt{2} K_0} \int_{r_0}^{x} \rho_{\Delta x} (r + j\Delta x, 0) \|\Delta \mathbf{z}\| r^2 \, dr$$

$$\leq \frac{\kappa F_2^*(G_2 \cdot t_j) \bar{S} x}{\sqrt{2} K_0} \int_{r_0}^{x} \|\Delta \mathbf{z}\| \, dr \tag{4.8.28}$$

$$\leq \frac{\kappa F_2^*(G_2 \cdot t_j) \bar{S} x^2}{\sqrt{2} K_0 L} G_2 |t_j - t_{j_0}|,$$

where we have used

$$\int_{r_0}^{x} \|\Delta \mathbf{z}\| \, dr \leq \frac{x}{L} G_2 |t_{j_2} - t_{j_1}|, \tag{4.8.29}$$

a consequence of (4.7.86). Note that the factor x/L bounds the number of intervals of length L between r_0 and x. We record this as a Corollary of Theorem 14:

Corollary 4 *Assume Corollary 3, Theorem 14, applies up to time T_0. Then*

$$|M_{\Delta x}(x, t_j) - M_{\Delta x}(x, t_{j_0})| \leq \frac{\kappa F_2^* (G_2 \cdot t_j) \bar{S} x^2}{\sqrt{2} K_0 L} G_2 |t_j - t_{j_0}|, \tag{4.8.30}$$

for all $0 \leq t_{j_0} \leq t_j \leq T_0$.

In particular, ignoring errors of order Δx, (4.8.30) implies the local Lipschitz in time continuity of $M_{\Delta x}$, (and hence of $B_{\Delta x}$ and $A_{\Delta x}$).

We can estimate $|M_{\Delta x}(x, t_j) - M_{\Delta x}(x, t_{j_0})|$ differently starting from (4.8.27) as follows:

$$
\begin{aligned}
&|M_{\Delta x}(x, t_j) - M_{\Delta x}(x, t_{j_0})| \\
&\leq \frac{\kappa F_2^* (G_2 \cdot t_j)}{\sqrt{2} K_0} \left[\int_{r_0}^R + \int_R^\infty \right] \rho_{\Delta x} \left(r + j\Delta x, 0 \right) \| \Delta \mathbf{z} \| r^2 \, dr, \\
&\leq \frac{\kappa F_2^* (G_2 \cdot t_j) \bar{S} R^2}{\sqrt{2} K_0 L} G_2 |t_j - t_{j_0}| \qquad\qquad (4.8.31) \\
&\quad + \frac{\kappa F_2^* (G_2 \cdot t_j) F_0^* (G_2 \cdot t_j)}{\sqrt{2} K_0} \int_R^\infty \rho_{\Delta x} \left(r + j\Delta x, 0 \right) r^2 \, dr,
\end{aligned}
$$

where we have used (4.8.28) together with

$$
\| \Delta \mathbf{z} \| \leq F_0^* (G_2 \cdot t_j), \qquad\qquad (4.8.32)
$$

a consequence of (4.7.88). But

$$
\begin{aligned}
\frac{\kappa}{2} \int_R^\infty \rho_{\Delta x} \left(r + j\Delta x, 0 \right) r^2 \, dr &\leq \frac{\kappa}{2} \int_R^\infty \rho_{\Delta x}(r + j\Delta x, 0)(r + j\Delta x)^2 \, dr \\
&\leq \frac{\kappa}{2} \int_R^\infty \rho_{\Delta x}(r, 0) r^2 \, dr \\
&\leq \frac{\kappa}{2} \int_R^\infty u_{\Delta x}^0(r, 0) r^2 \, dr \\
&\leq M_{\Delta x}(\infty, 0) - M_{\Delta x}(R, 0), \qquad (4.8.33)
\end{aligned}
$$

and since

$$
\lim_{R \to \infty} [M_{\Delta x}(R, 0) - M_{\Delta x}(\infty, 0)] = 0, \qquad\qquad (4.8.34)
$$

it follows that for any $\delta > 0$ sufficiently small, there exists $R(\delta) > 0$, such that

$$
\frac{\kappa F_2^* (G_2 \cdot t_j) F_0^* (G_2 \cdot t_j)}{\sqrt{2} K_0} \int_{R(\delta)}^\infty \rho_{\Delta x} \left(r + j\Delta x, 0 \right) r^2 \, dr \leq \delta, \qquad (4.8.35)
$$

for all $t_j \leq T_0$. Indeed, since $M_{\Delta x}(x, 0)$ is a continuous monotone increasing function of x, it follows that we can define $R(\delta)$ to satisfy the equality

$$
\frac{\sqrt{2}}{K_0} F_2^* (G_2) F_0^* (G_2) [M_{\Delta x}(\infty, 0) - M_{\Delta x}(R(\delta), 0)] = \delta, \qquad (4.8.36)
$$

in which case (4.8.35) follows at once from (4.8.33). Using this definition of $R(\delta)$ in (4.8.31), it follows that for every $\delta > 0$,

$$|M_{\Delta x}(x,t_j) - M_{\Delta x}(x,t_{j_0})| \leq \frac{\kappa F_2^* (G_2 \cdot t_j) \bar{S}R(\delta)^2}{\sqrt{2}K_0 L}G_2|t_j - t_{j_0}| + \delta. \quad (4.8.37)$$

Therefore, assuming Corollary 3, Theorem 14 applies up to some time T_0, $0 < T_0 \leq 1$, we can choose $\delta = \varepsilon/2$, and set

$$T_\varepsilon = \left\{ \frac{\kappa F_2^* (G_2) \bar{S}R(\varepsilon/2)^2}{\sqrt{2}K_0 L}G_2 \right\}^{-1} \frac{\varepsilon}{2}, \quad (4.8.38)$$

and conclude from (4.8.37) that

$$|M_{\Delta x}(x,t_j) - M_{\Delta x}(x,t_{j_0})| < \varepsilon, \quad (4.8.39)$$

for all $t_j \leq Max\{T_\varepsilon, T_0\}$. We record this as another corollary to Theorem 14:

Corollary 5 *Assume that Corollary 3, Theorem 14, holds up to time T_0. Then for all $\varepsilon > 0$, there exists $T_\varepsilon > 0$, (given explicitly in (4.8.38)), such that*

$$|M_{\Delta x}(x,t_j) - M_{\Delta x}(x,t_{j_0})| < \varepsilon, \quad (4.8.40)$$

for all $x \geq r_0$, $t_j \leq Min\{T_\varepsilon, T_0\}$.

We now use Corollary 5 to define $\bar{B}$ and T_B. Consider the function $B_{\Delta x}(x,t)$. Assume that the initial data satisfies

$$B_{\Delta x}(x,0) = \frac{1}{1 - \frac{2M_{\Delta x}(x,0)}{x}} \leq \bar{B}_0 \quad (4.8.41)$$

for some positive constant $\bar{B}_0$. Choose $\bar{B} > \bar{B}_0$, say

$$\bar{B} = 2\bar{B}_0. \quad (4.8.42)$$

Choose $\varepsilon > 0$ by

$$\varepsilon = Sup \left\{ \varepsilon : \frac{1}{1 - \frac{2(M_{\Delta x}(x,0)+\varepsilon)}{x}} \leq 2\bar{B}_0 = \bar{B}, \ for \ all \ r_0 \leq x < \infty \right\}. \quad (4.8.43)$$

Claim: By (4.8.43),

$$\varepsilon \geq \frac{r_0}{2} \left(\frac{1}{\bar{B}_0} - \frac{1}{2\bar{B}_0} \right) > 0. \quad (4.8.44)$$

To see this, let $\varepsilon(x)$ be defined so that

$$\frac{1}{1 - \frac{2(M_{\Delta x}(x,0) + \varepsilon(x))}{x}} = 2\bar{B}_0. \tag{4.8.45}$$

Solving (4.8.45) for $\varepsilon(x)$ gives

$$\varepsilon(x) = \frac{x}{2}\left\{1 - \frac{1}{2\bar{B}_0} - \frac{2M_{\Delta x}(x,0)}{x}\right\}. \tag{4.8.46}$$

But (4.8.41) implies

$$\frac{2M_{\Delta x}(x,0)}{x} \leq 1 - \frac{1}{\bar{B}_0}. \tag{4.8.47}$$

Using (4.8.47) in (4.8.46) gives (4.8.44). $\square$

Now for ε in (4.8.43), define

$$T_B = T_\varepsilon \equiv \left\{\frac{\kappa F_2^*\left(G_2\right)\bar{S}R(\varepsilon/2)^2}{\sqrt{2}K_0 L}G_2\right\}^{-1}\frac{\varepsilon}{2}, \tag{4.8.48}$$

so that by (4.8.38), (4.8.40),

$$|M_{\Delta x}(x, t_j) - M_{\Delta x}(x, 0)| < \varepsilon, \tag{4.8.49}$$

for all $t_j \leq Max\{T_\varepsilon, T_0\}$. But (4.8.40), (4.8.43), directly imply

$$\frac{1}{1 - \frac{2(M_{\Delta x}(x,0) + \varepsilon)}{x}} \leq \bar{B}. \tag{4.8.50}$$

We conclude that assumption (4.7.30) of Corollary 3, Theorem 14, can be replaced by the condition that $\bar{B}$ is defined in (4.8.42), together with the condition that $t_j \leq T_B$, where T_B is defined in (4.8.48). We have shown that assumptions (4.7.29)-(4.7.32) of Corollary 3, Theorem 14, can be removed, and are consequences of appropriately restricting the time T_0 and redefining the constants involved in terms of the initial data.

The following theorem, which summarizes our results, follows directly from our construction of $\bar{v}, \bar{S}, \bar{M}, \bar{B}$ and T_S, T_M, T_B above:

Theorem 15 *Let $u_{\Delta x}(x,t)$, $\mathbf{A}_{\Delta x}(x,t)$ be an approximate solution generated by the fractional step Glimm method starting from initial data $u_{\Delta x}(x,0)$, $\mathbf{A}_{\Delta x}(x,0)$, and let $\bar{M}_0, \bar{B}_0, \bar{S}_0, \bar{v}_0$ and $\bar{V}_0$ be positive constants such that the initial data satisfies:*

$$M_{\Delta x}(x,0) \leq \bar{M}_0, \tag{4.8.51}$$
$$B_{\Delta x}(x,0) \leq \bar{B}_0, \tag{4.8.52}$$
$$0 < S_{\Delta x}(x,0) \leq \bar{S}_0 \tag{4.8.53}$$
$$|v_{\Delta x}(x,0)| \leq \bar{v}_0 < c, \tag{4.8.54}$$

for all $x \geq r_0$, and

$$\sum_{i_1 \leq i \leq i_2, \ p=1,2} |\gamma_{i0}^p| < V_0, \tag{4.8.55}$$

for all $r_0 \leq x_{i_2} < x_{i_2} < \infty$, $|x_{i_2} - x_{i_1}| \leq L$. Let $\bar{v} = 2\bar{v}_0$, $\bar{S} = 2\bar{S}_0$, $\bar{M} = 2\bar{M}_0$, $\bar{B} = 2\bar{B}_0$, assume that

$$\frac{\Delta x}{\Delta t} = \Lambda = 2\sqrt{G_{AB}(\bar{B}, \bar{M})}, \tag{4.8.56}$$

and let

$$T = Min\{1, T_2, T_{\bar{S}}, T_{\bar{M}}, T_{\bar{B}}\}, \tag{4.8.57}$$

where

$$T_2 = T_2 = \left(\frac{1}{G_2}\right) \frac{\bar{V}_*}{\{2\bar{V}_* + H(2\bar{V}_*)\}}, \tag{4.8.58}$$

$$T_S = Sup\{t : F_1^*(G_2 \cdot t_j) \leq 2F_1^*(0), \quad \text{for all } t_j \leq t\},$$

$$T_M = T_M = Sup\{t : M_{r_0} + F_2^*(G_2 \cdot t_j) M_0 \leq \bar{M}, \text{ for all } t_j \leq t\},$$

$$T_B = \left\{\frac{\kappa F_2^*(G_2) \bar{S} R(\varepsilon/2)^2}{\sqrt{2} K_0 L} G_2\right\}^{-1} \frac{\varepsilon}{2},$$

and

$$\varepsilon = Sup\left\{\varepsilon : \frac{1}{1 - \frac{2(M_{\Delta x}(x,0) + \varepsilon)}{x}} \leq \bar{B}, \text{ for all } r_0 \leq x < \infty\right\}, \tag{4.8.59}$$

c.f., (4.7.84), (4.8.10), (4.8.18), (4.8.48) and (4.8.43). Then the approximate solution $u_{\Delta x}$, $\mathbf{A}_{\Delta x}$ is well defined for all $r_0 \leq r < \infty$, $0 \leq t \leq T$, and satisfies the bounds

$$M_{\Delta x}(x, t_j) \leq \bar{M}, \tag{4.8.60}$$

$$B_{\Delta x}(x, t_j) \leq \bar{B}, \tag{4.8.61}$$

$$0 < S_{\Delta x}(x, t_j) \leq \bar{S} \tag{4.8.62}$$

$$|v_{\Delta x}(x, t_j)| \leq \bar{v} < c, \tag{4.8.63}$$

together with the bounds

$$\sum_{i_1 \leq i \leq i_2, \ p=1,2} |\gamma_{ij}^p| < 2\bar{V}_* = 2\left(1 + \frac{4\sqrt{G_{AB}}}{L}\right) V_0, \tag{4.8.64}$$

$$\|\mathbf{z}_{ij} - \mathbf{z}_{i+j,0}\| \leq F_0^*(G_2 \cdot T),$$ (4.8.65)

$$\int_{x_{i_1}}^{x_{i_2}} \|\mathbf{z}_{\Delta x}(x, t_{j_2}) - \mathbf{z}_{\Delta x}(x, t_{j_1})\| \, dx \leq G_2 |t_{j_2} - t_{j_1}|,$$ (4.8.66)

$$|A'_{\Delta x}(x, t_j)| \leq \left(\frac{1}{r_0} + \kappa \frac{c^2 + \sigma^2 \bar{v}^2}{c^2 - \bar{v}^2} \bar{S} \right) \bar{B} G_{AB},$$ (4.8.67)

$$|B'_{\Delta x}(x, t_j)| \leq \left(\frac{1}{r_0} + \kappa \frac{c^2 + \sigma^2 \bar{v}^2}{c^2 - \bar{v}^2} \bar{S} \right) \bar{B}^2,$$ (4.8.68)

$$|M_{\Delta x}(x, t_{j_2}) - M_{\Delta x}(x, t_{j_1})| \leq \frac{\kappa F_2^* (G_2 \cdot T) \bar{S} x^2}{\sqrt{2} K_0 L} G_2 |t_{j_2} - t_{j_1}|,$$ (4.8.69)

for all $r_0 \leq x, x_{i_1}, x_{i_2} < \infty$, $|x_{i_2} - x_{i_1}| \leq L$, and $0 \leq t_j, t_{j_1}, t_{j_2} \leq T$, c.f. (4.7.83), (4.7.90), (4.7.41), (4.8.30), (4.7.86), (4.7.43), (4.7.44).

Recall that the constants $G_{AB} \equiv G_{AB}(\bar{M}, \bar{B})$, $G_2 \equiv G_2(\bar{M}, \bar{B}, \bar{S})$, $F_1^*(G_2 \cdot T)$, $F_2^*(G_2 \cdot T)$, and $V_*(G_2 \cdot T)$, defined in (4.3.43),(4.7.81),(4.8.7),(4.8.15), and (4.7.85) respectively, are based on the functions $G_{AB}(\cdot, \cdot)$, $G_2(\cdot, \cdot, \cdot)$, $F_i^*(\cdot)$ and $V_*(\cdot)$ that depend only on the constants σ, c, K_0, r_0, L and V_0, and thus are determined by the initial data alone.

Corollary 6 *Let $u_{\Delta x}(x, t)$, $\mathbf{A}_{\Delta x}(x, t)$ be an approximate solution generated by the fractional step Glimm method starting from initial data $u_{\Delta x}(x, 0)$, $\mathbf{A}_{\Delta x}(x, 0)$, that satisfies the conditions (4.8.60)-(4.8.59) of Theorem 15. Then there exists a subsequence $\Delta x \to 0$ and bounded measurable functions $u(x, t) = \Psi^{-1} \cdot \Phi \cdot \mathbf{z}(x, t)$, $\mathbf{A}(x, t)$, such that $(u_{\Delta x}, \mathbf{A}_{\Delta x}) \to (u, \mathbf{A})$ for a.e. $(x, t) \in [r_0, \infty) \times [0, T]$. Moreover, the convergence $u_{\Delta x}(\cdot, t) \to u(\cdot, t)$ is in L^1_{loc} for each $t \in [0, T]$, uniformly on compact sets in (x, t)-space, and the limit function $u_{\Delta x}$ satisfies:*

$$TV_{[x_1, x_2]} z(\cdot, t) \leq 2\bar{V}_*,$$
$$TV_{[x_1, x_2]} w(\cdot, t) \leq H(2\bar{V}_*),$$ (4.8.70)
$$TV_{[x_1, x_2]} \mathbf{z}(\cdot, t) \leq 2\bar{V}_* + H(2\bar{V}_*),$$

$$\|\mathbf{z}(x, t) - \mathbf{z}(x + \lambda T, 0)\| \leq F_0^*(G_2 \cdot T),$$ (4.8.71)

and

$$\int_{x_1}^{x_2} \|\mathbf{z}(x, t_2) - \mathbf{z}(x, t_1)\| \, dx \leq G_2 |t_2 - t_1|,$$ (4.8.72)

for all $r_0 \leq x, x_1, x_2 < \infty$, $|x_2 - x_1| < L$, and $0 \leq t, t_1, t_2 \leq T$.

The convergence in **A** *is pointwise a.e., uniformly on compact sets in* (x,t)-*space, and the limit function* **A**(x,t) *satisfies*

$$A(x,t) = A_{r_0} \exp \int_{r_0}^{x} \left\{ \frac{B(r,t)-1}{r} + \kappa r B(r,t) T_M^{11}(u(r,t)) \right\} dr, \qquad (4.8.73)$$

$$B(r,t) = \frac{1}{1 - \frac{2M(r,t)}{r}}, \quad M(r,t) = M(r_0,t) + \frac{\kappa}{2} \int_{r_0}^{r} u^0(r,t) r^2 \, dr, \qquad (4.8.74)$$

$$\left| \frac{A(x+y,t) - A(x,t)}{y} \right| \le \left(\frac{1}{r_0} + \kappa \frac{c^2 + \sigma^2 \bar{v}^2}{c^2 - \bar{v}^2} \bar{S} \right) \bar{B} G_{AB}, \qquad (4.8.75)$$

$$\left| \frac{B(x+y,t) - B(x,t)}{y} \right| \le \left(\frac{1}{r_0} + \kappa \frac{c^2 + \sigma^2 \bar{v}^2}{c^2 - \bar{v}^2} \bar{S} \right) \bar{B}^2, \qquad (4.8.76)$$

$$|M(x,t_2) - M(x,t_1)| \le \frac{\kappa F_2^* (G_2 \cdot T) \bar{S} x^2}{\sqrt{2} K_0 L} G_2 |t_2 - t_1|, \qquad (4.8.77)$$

for all $r_0 \le x, x_1, x_2 < \infty$, $|x_2 - x_1| \le L$, *and* $0 \le t, t_1, t_2 \le T$.

Proof: It follows from (4.7.90), (4.8.64) (together with the non-singularity of the mapping from **z** $\to u$) that the approximate solution $u_{\Delta x}(x,t)$ is bounded, and of locally bounded total variation at each fixed time $0 \le t \le T$, and these bounds are uniform in time over compact x-intervals. Moreover, it follows from (4.8.66) that $u_{\Delta x}(x,t)$ is locally Lipschitz continuous in the L^1-norm at each time, uniformly on compact sets. These bounds are uniform as $\Delta x \to 0$. This is all that is required to apply Oleinik's compactness argument to the function $u_{\Delta x}$, [10, 26, 19]. From this we can conclude that there exists a sequence $\Delta x \to 0$ such that $u_{\Delta x}$ converges a.e. to a bounded measurable function u on $x \ge r_0, 0 \le t \le T$. The convergence is in L^1_{loc} at each time, uniformly on compact sets, and the supnorm bound (4.8.71), the local total variation estimate (4.8.70), and the continuity of the local L^1 norm (4.8.72), carry over from the corresponding estimates (4.8.71), (4.8.64), (4.7.86) for approximate solution. (For (4.8.64) we use that the change Δw across a wave is bounded by $H(\Delta z)$, and that H is a convex function, c.f. Proposition []. The Oleinik argument is based on using Helly's Theorem to extract a pointwise convergent subsequence on a dense set of times between $t = 0$ and $T = t$, and then to use the local L^1-Lipschitz continuity of $u_{\Delta x}$ to extrapolate the L^1 convergence to all intermediate times, [19].)

It follows from (4.8.69)-(4.8.68), together with (4.3.28), that $\mathbf{A}_{\Delta x}$ is locally Lispchitz continuous in x and t for $x \ge r_0$, $t \le T$, (ignoring errors that are of order Δx), and the Lipschitz bounds are uniform as $\Delta x \to 0$. It follows from Arzela-Ascoli that on some subsequence $\Delta x \to 0$, $\mathbf{A}_{\Delta x}$ converges to a locally Lipschitz continuous function $\mathbf{A}(x,t)$, and the convergence is pointwise almost everywhere, uniformly on compact sets. It follows that the convergence of $u_{\Delta x}$ and $\mathbf{A}_{\Delta x}$ is strong enough to pass the limit through the integral sign in (4.3.28) and (1.3.15), and thus conclude (4.8.73) and (4.8.74), respectively. Similarly, (4.8.76)-(4.8.77) are obtained from (4.7.43)-(4.8.30), respectively. The initial data u_0 is taken on in the L^1 sense,

$$\lim_{t \to 0} \|u(\cdot, t) - u_0(\cdot)\|_{L^1_{loc}} = 0, \tag{4.8.78}$$

and the boundary condition $v = 0 \iff M(r_0, 0) = M_{r_0}$ is taken on weakly, c.f. [19].$\square$

Proof of Theorem 13: In the final section we prove that for almost every sample sequence $\mathbf{a}$, the functions $u_{\Delta x}(x,t), \mathbf{A}(x,t)$ define a weak solution of the Einstein equations (1.3.2)-(1.3.5) on $r_0 \le x < \infty, 0 \le t \le T$. Assuming this, $u_{\Delta x}(x,t), \mathbf{A}(x,t)$ is then a weak solution of (1.3.2)-(1.3.5) in the class $u_{\Delta x}$ bounded measurable and $\mathbf{A}_{\Delta x}$ Lipschitz continuous, and so it follows that our results in [12] apply. In particular, (1.3.3) holds in the pointwise almost everywhere sense. Thus the proof of Theorem 13 is complete once we verify (4.2.42). (The assumptions (4.2.28)-(4.2.29) just imply that $TV_{[x_1,x_2]}\mathbf{z}(\cdot, 0) < \infty$, and this guarantees (4.8.64).) For (4.2.42), note first that (4.1.3) together with (4.8.71) imply that

$$\lim_{x \to 0} \dot{M}(x,t) = 0, \tag{4.8.79}$$

for all $0 \le t \le T$. To see this, recall from Theorem 12 that if $u_{\Delta x}, \mathbf{A}_{\Delta x}$ is a weak solution for $0 \le t \le T$, then (1.3.3) and (1.3.8) hold. By (1.3.8), statement (4.8.79) follows so long as

$$\lim_{r \to \infty} \sqrt{\frac{A(r,t)}{B(r,t)}} u^1(r,t)r^2 = 0 \tag{4.8.80}$$

for $t \le T$, where

$$\left| \sqrt{\frac{A(r,t)}{B(r,t)}} u^1(r,t)r^2 \right| \le \sqrt{A(r,t)} u^0(r,t)r^2. \tag{4.8.81}$$

Now since A and B are given by (4.8.73) and (4.8.74), it follows that $\mathbf{A}$ satisfies (1.3.2) and (1.3.4), and so adding these two equations, and following the argument leading to (4.3.43), we obtain that

$$|A| \le A_{r_0} B_{r_0} exp\left\{ \frac{8\bar{B}\bar{M}}{r_0} \right\},$$

and thus A is uniformly bounded. Since $|v(x,t)| \le \bar{v} < c$, (4.1.3) and (4.8.71) imply that

$$\lim_{r\to\infty} \sqrt{A(r,t)} u^0(r,t) r^2 = 0, \tag{4.8.82}$$

and so (4.8.79) follows as claimed. But (4.8.79) implies that,

$$\lim_{x\to\infty} M(x,t) = \lim_{x\to\infty} M(x,0) = M_\infty, \tag{4.8.83}$$

which is (4.2.42) of Theorem 13. We conclude from Theorem 12 that the proof of Theorem 13 is complete once we prove that $u(x,t)$, $\mathbf{A}(x,t)$ is a genuine weak solution of (4.1.5),(4.1.6) with initial boundary data 4.2.36)-(4.2.38). This is the topic of the next section. $\square$

Our theorems have the following corollary:

Corollary 7 *Assume that the initial data $u_0(x)$ satisfies (1)-(5). Then a bounded weak solution $u(x,t)$, $\mathbf{A}(x,t)$ of the Einstein equations (1.3.2)-(1.3.5) exists up until the first time T at which either*

$$\lim_{t\to T^-} Sup_x B(x,t) = \infty, \tag{4.8.84}$$

$$\lim_{t\to T^-} Sup_x x\rho(x,t) = \infty, \tag{4.8.85}$$

or

$$\lim_{t\to T^-} Sup_x TV_{[x_1,x_2]}\mathbf{z}(\cdot,t) = \infty. \tag{4.8.86}$$

Proof: If B, S and $TV_{[x_1,x_2]}\mathbf{z}$ remain uniformly bounded up to time T, then our argument shows that v remains uniformly bounded away from c up to time T, c.f. (4.8.2)-(4.8.3). Thus we can repeat the proof that the solution starting from initial data at time T, continues forward for some positive time. The Corollary follows at once.

4.9 Convergence

In this section we prove that the limits u, $\mathbf{A}$ of approximate solutions $u_{\Delta x}$, $\mathbf{A}_{\Delta x}$ established in Corollary 6, are weak solutions of (1.4.3), (1.4.4), for almost every choice of sample sequence $\mathbf{a}$. This is a modification of Glimm's original argument [10], as well as the argument in [19]. The main point is to show that the the discontinuities in $\mathbf{A}$ at the boundary of the mesh rectangles $\mathcal{R}_{ij}$ are accounted for by inclusion of the term

$$\mathbf{A}' \cdot \nabla_{\mathbf{A}} f(\mathbf{A}, u, x) = \frac{1}{2}\sqrt{\frac{A}{B}} \delta\left(T_M^{01}, T_M^{11}\right),$$

in the ODE step (4.3.22), c.f. (4.3.17).

To start, recall that both u, $\mathbf{A}$ and $u_{\Delta x}$, $\mathbf{A}_{\Delta x}$ satisfy the estimates (4.8.70)-(4.8.77), and recall that $u_{\Delta x}^{RP}$ denotes the exact Riemann problem solution in each $\mathcal{R}_{ij}$ for the homogeneous system (4.4.1), so that

$$
\begin{aligned}
0 = \int\!\!\int_{\mathcal{R}_{ij}} & \{-u_{\Delta x}^{RP}\varphi_t - f(\mathbf{A}_{ij}, u_{\Delta x}^{RP})\varphi_x\}\, dxdt \\
+ & \int_{\mathcal{R}_i} \{u_{\Delta x}^{RP}(x, t_{j+1}^-)\varphi(x, t_{j+1}) - u_{\Delta x}^{RP}(x, t_j^+)\varphi(x, t_j)\}\, dx \quad (4.9.1)\\
+ & \int_{\mathcal{R}_j} \Big\{ f(\mathbf{A}_{ij}, u_{\Delta x}^{RP}(x_{i+\frac{1}{2}}, t))\varphi(x_{i+\frac{1}{2}}, t) \\
& - f(\mathbf{A}_{ij}, u_{\Delta x}^{RP}(x_{i-\frac{1}{2}}, t))\varphi(x_{i-\frac{1}{2}}, t) \Big\}\, dt.
\end{aligned}
$$

Recall also that $\hat{u}(t, u_0)$ denotes the solution to the initial value problem

$$
\begin{aligned}
\hat{u}_t &= G(\mathbf{A}_{ij}, \hat{u}, x) = g(\mathbf{A}_{ij}, \hat{u}, x) - \mathbf{A}' \cdot \nabla_{\mathbf{A}} f(\mathbf{A}_{ij}, \hat{u}, x), \\
\hat{u}(0) &= u_0.
\end{aligned}
$$

Thus

$$
\begin{aligned}
\hat{u}(t, u_0) - u_0 &= \int_0^t \hat{u}_t\, dt \\
&= \int_0^t \{g(\mathbf{A}_{ij}, \hat{u}(\xi, u_0), x) - \mathbf{A}' \cdot \nabla_{\mathbf{A}} f(\mathbf{A}_{ij}, \hat{u}, x)\}\, dt.
\end{aligned}
$$

Since $\hat{u}$ implements the ODE step of the fractional step method, it follows that the approximate solution $u_{\Delta x}(x, t)$ is defined on each mesh rectangle $\mathcal{R}_{ij}$ by the formula

$$
\begin{aligned}
u_{\Delta x}(x, t) = u_{\Delta x}^{RP}(x, t) &+ \int_{t_j}^t \{g(\mathbf{A}_{ij}, \hat{u}(\xi - t_j, u_{\Delta x}^{RP}(x, t)), x)\}\, dt \quad (4.9.2)\\
&- \int_{t_j}^t \Big\{ \frac{\partial f}{\partial \mathbf{A}}(\mathbf{A}_{ij}, \hat{u}(\xi - t_j, u_{\Delta x}^{RP}(x, t))) \cdot \mathbf{A}'_{\Delta x} \Big\}\, dt.
\end{aligned}
$$

Note that the difference between the approximate and Riemann problem solutions is on the order of Δx. Define the residual $\varepsilon(u_{\Delta x}, \mathbf{A}_{\Delta x}, \varphi)$ of the approximate solutions $u_{\Delta x}$ by

$$
\begin{aligned}
\varepsilon(u_{\Delta x}, \mathbf{A}_{\Delta x}, \varphi) = \int_{r_0}^\infty \int_0^\infty & \{-u_{\Delta x}\varphi_t - f(\mathbf{A}_{\Delta x}, u_{\Delta x})\varphi_x \\
& - g(\mathbf{A}_{\Delta x}, u_{\Delta x}, x)\varphi\}\, dtdx - I_1 - I_2,
\end{aligned}
$$

$$= \sum_{ij} \int\int_{\mathcal{R}_{ij}} \{ -u_{\Delta x}\varphi_t - f(\mathbf{A}_{ij}, u_{\Delta x})\varphi_x$$

$$-g(\mathbf{A}_{\Delta x}, u_{\Delta x}, x)\varphi\} \, dtdx - I_1 - I_2, \quad (4.9.3)$$

where

$$I_1 = \int_{r_0}^{\infty} u_{\Delta x}(x, 0^+)\varphi(x, 0) \, dx$$

$$= \sum_i \int_{\mathcal{R}_i} u_{\Delta x}(x, 0^+)\varphi(x, 0) \, dx, \quad (4.9.4)$$

and

$$I_2 = \int_0^{\infty} f(\mathbf{A}_{\Delta x}(r_0^+, t), u_{\Delta x}(r_0^+, t))\varphi(r_0, t) \, dt$$

$$= \sum_j \int_{\mathcal{R}_j} f(\mathbf{A}_{ij}, u_{\Delta x}(r_0^+, t))\varphi(r_0, t) \, dt. \quad (4.9.5)$$

We now prove that the residual is $O(\Delta x)$. (It follows that if $u_{\Delta x} \to u$ and $\mathbf{A}_{\Delta x} \to \mathbf{A}$ converge in L^1_{loc} at each time, uniformly on compact sets, then the limit function will satisfy $\varepsilon(u, \mathbf{A}, \varphi) = 0$, the condition that u be a weak solution of the Einstein equations.) To this end, substitute (4.9.2) into (4.9.3) to obtain

$$\varepsilon(u_{\Delta x}, \mathbf{A}_{\Delta x}, \varphi) = \sum_{ij} \int\int_{\mathcal{R}_{ij}} \{ -u_{\Delta x}^{RP}\varphi_t - f(\mathbf{A}_{ij}, u_{\Delta x})\varphi_x - g(\mathbf{A}_{ij}, u_{\Delta x}, x)\varphi$$

$$-\varphi_t \int_{t_j}^t [g(\mathbf{A}_{ij}, \hat{u}(\xi - t_j, u_{\Delta x}^{RP}(x, t)), x) \quad (4.9.6)$$

$$-\frac{\partial f}{\partial \mathbf{A}}(\mathbf{A}_{ij}, \hat{u}(\xi - t_j, u_{\Delta x}^{RP}(x, t))) \cdot \mathbf{A}'_{\Delta x} \Big] d\xi \} \, dxdt$$

$$-I_1 - I_2.$$

Set

$$I^1_{ij}(x, t) = \int_{t_j}^t [g(\mathbf{A}_{ij}, \hat{u}(\xi - t_j, u_{\Delta x}^{RP}(x, t)), x)] \, d\xi$$

$$\int_{t_j}^t \left[-\frac{\partial f}{\partial \mathbf{A}}(\mathbf{A}_{ij}, \hat{u}(\xi - t_j, u_{\Delta x}^{RP}(x, t))) \cdot \mathbf{A}'_{\Delta x} \right] d\xi.$$

Upon substituting (4.9.1) into (4.9.7), we have

$$\varepsilon(u_{\Delta x}, \mathbf{A}_{\Delta x}, \varphi) = \sum_{ij} \int\int_{\mathcal{R}_{ij}} \left\{ \varphi_x \left[f(\mathbf{A}_{ij}, u_{\Delta x}^{RP}) - f(\mathbf{A}_{ij}, u_{\Delta x}) \right] \right.$$

$$\left. -g(\mathbf{A}_{ij}, u_{\Delta x}, x)\varphi - \varphi_t I_{ij}^1(x,t) \right\} dxdt \qquad (4.9.7)$$

$$-I_1 - \sum_{ij} \int_{\mathcal{R}_i} \left\{ u_{\Delta x}^{RP}(x, t_{j+1}^-)\varphi(x, t_{j+1}) - u_{\Delta x}^{RP}(x, t_j^+)\varphi(x, t_j) \right\} dx$$

$$-I_2 - \sum_{ij} \int_{\mathcal{R}_j} \left\{ f(\mathbf{A}_{ij}, u_{\Delta x}^{RP}(x_{i+\frac{1}{2}}, t))\varphi(x_{i+\frac{1}{2}}, t) \right.$$

$$\left. -f(\mathbf{A}_{ij}, u_{\Delta x}^{RP}(x_{i-\frac{1}{2}}, t))\varphi(x_{i-\frac{1}{2}}, t) \right\} dt$$

Note that

$$|f(\mathbf{A}_{ij}, u_{\Delta x}^{RP}) - f(\mathbf{A}_{ij}, u_{\Delta x})| \le C\Delta t, \qquad (4.9.8)$$

and so

$$\left| \sum_{ij} \int\int_{\mathcal{R}_{ij}} \varphi \left[f(\mathbf{A}_{ij}, u_{\Delta x}^{RP}) - f(\mathbf{A}_{ij}, u_{\Delta x}) \right] dxdt \right| \le |\varphi_x|_\infty C\Delta t T(b-a),$$

$$(4.9.9)$$

where $Supp(\varphi) \subset [a,b] \times [0.T]$. (We let C denote a generic constant that depends only on the bounds for the solution.) Using the fact that $u_{\Delta x}^{RP}(x, t_j^+) = u_{\Delta x}(x, t_j^+)$, and inserting (4.9.2), we obtain that

$$-I_1 - \sum_{ij} \int_{\mathcal{R}_i} \left\{ u_{\Delta x}^{RP}(x, t_{j+1}^-)\varphi(x, t_{j+1}) - u_{\Delta x}^{RP}(x, t_j^+)\varphi(x, t_j) \right\} dx$$

$$= \sum_{j \ne 0} \int_{r_0}^\infty \left\{ u_{\Delta x}(x, t_j^+) - u_{\Delta x}^{RP}(x, t_j^-) \right\} \varphi(x, t_j) \, dx$$

$$= \sum_{j \ne 0} \int_{r_0}^\infty \varphi(x, t_j) \left\{ u_{\Delta x}(x, t_j^+) - u_{\Delta x}(x, t_j^-) \right\} dx \qquad (4.9.10)$$

$$+ \sum_{j \ne 0} \int_{r_0}^\infty \varphi(x, t_j) \left\{ u_{\Delta x}(x, t_j^-) - u_{\Delta x}^{RP}(x, t_j^-) \right\} dx.$$

Set

$$\varepsilon_1(u_{\Delta x}, \mathbf{A}_{\Delta x}, \varphi) = \sum_{j \ne 0} \int_{r_0}^\infty \varphi(x, t_j) \left\{ u_{\Delta x}(x, t_j^+) - u_{\Delta x}(x, t_j^-) \right\} dx. \quad (4.9.11)$$

It follows that

$$\varepsilon(u_{\Delta x}, \mathbf{A}_{\Delta x}, \varphi) = O(\Delta x) + \varepsilon_1(u_{\Delta x}, \mathbf{A}_{\Delta x}, \varphi) \tag{4.9.12}$$

$$+ \sum_{ij} \int\int_{\mathcal{R}_{ij}} \left\{ -g(\mathbf{A}_{ij}, u_{\Delta x}, x)\varphi - \varphi_t I_{ij}^1(x,t) \right\} dx dt$$

$$+ \sum_{j \neq 0} \int_{r_0}^{\infty} \varphi(x, t_j) \left\{ u_{\Delta x}(x, t_j^-) - u_{\Delta x}^{RP}(x, t_j^-) \right\} dx$$

$$- I_2 - \sum_{ij} \int_{\mathcal{R}_j} \left\{ f(\mathbf{A}_{ij}, u_{\Delta x}^{RP}(x_{i+\frac{1}{2}}, t))\varphi(x_{i+\frac{1}{2}}, t) \right.$$

$$\left. - f(\mathbf{A}_{ij}, u_{\Delta x}^{RP}(x_{i-\frac{1}{2}}, t))\varphi(x_{i-\frac{1}{2}}, t) \right\} dt.$$

But

$$- I_2 - \sum_{ij} \int_{\mathcal{R}_j} \left\{ f(\mathbf{A}_{ij}, u_{\Delta x}^{RP}(x_{i+\frac{1}{2}}, t))\varphi(x_{i+\frac{1}{2}}, t) \right.$$

$$\left. - f(\mathbf{A}_{ij}, u_{\Delta x}^{RP}(x_{i-\frac{1}{2}}, t))\varphi(x_{i-\frac{1}{2}}, t) \right\} dt$$

$$= \sum_{ij} \int_{\mathcal{R}_j} \left\{ f(\mathbf{A}_{i+1,j}, u_{\Delta x}^{RP}(x_{i+\frac{1}{2}}, t)) - f(\mathbf{A}_{ij}, u_{\Delta x}^{RP}(x_{i+\frac{1}{2}}, t)) \right\} \varphi(x_{i+\frac{1}{2}}, t) dt$$

$$+ \sum_{j} \int_{\mathcal{R}_j} \left\{ f(\mathbf{A}_{0j}, u_{\Delta x}^{RP}(r_0^+, t)) - f(\mathbf{A}_{0j}, u_{\Delta x}(r_0^+, t)) \right\} \varphi(r_0, t) dt,$$

$$\tag{4.9.13}$$

where

$$\left| \sum_{j} \int_{\mathcal{R}_j} \left\{ f(\mathbf{A}_{0j}, u_{\Delta x}^{RP}(r_0^+, t)) - f(\mathbf{A}_{0j}, u_{\Delta x}(r_0^+, t)) \right\} \varphi(r_0, t) dt \right|$$

$$\leq |\varphi|_\infty C \Delta t^2 \left(\frac{T}{\Delta t} \right) = O(\Delta t). \tag{4.9.14}$$

To analyze the term multiplied by φ_t in (4.9.12), we add and subtract a term that differs from this one by $O(\Delta x)$, and then use integration by parts on the new term. That is, set

$$I_{\Delta S} = \sum_{ij} \int\int_{\mathcal{R}_{ij}} \varphi_t \int_{t_j}^{t} \left[g(\mathbf{A}_{ij}, \hat{u}(\xi - t_j, u_{\Delta x}^{RP}(x, \xi)), x) \right.$$

$$- g(\mathbf{A}_{ij}, \hat{u}(\xi - t_j, u_{\Delta x}^{RP}(x, t)), x)$$

$$- \frac{\partial f}{\partial \mathbf{A}}(\mathbf{A}_{ij}, \hat{u}(\xi - t_j, u_{\Delta x}^{RP}(x, \xi))) \cdot \mathbf{A}'_{\Delta x}$$

$$\left. + \frac{\partial f}{\partial \mathbf{A}}(\mathbf{A}_{ij}, \hat{u}(\xi - t_j, u_{\Delta x}^{RP}(x, t))) \cdot \mathbf{A}'_{\Delta x} \right] d\xi \, dx dt.$$

But

$$|I_{\Delta S}| \le \sum_{ij} \int \int_{\mathcal{R}_{ij}} |\varphi_t|_\infty \int_{t_j}^t C|\gamma_{ij}^l| \, d\xi \, dx dt \le |\varphi_t|_\infty C \Delta t^2 \Delta x \sum_{ijl} |\gamma_{ij}^l|$$

$$\le CV|\varphi_t|_\infty \Delta t^2 \Delta x \frac{T}{\Delta t} = O(\Delta x^2),$$

and so

$$-\sum_{ij} \int \int_{\mathcal{R}_{ij}} \varphi_t \int_{t_j}^t \left[g(\mathbf{A}_{ij}, \hat{u}(\xi - t_j, u_{\Delta x}^{RP}(x,t)), x) \right.$$

$$\left. - \frac{\partial f}{\partial \mathbf{A}}(\mathbf{A}_{ij}, \hat{u}(\xi - t_j, u_{\Delta x}^{RP}(x,t))) \cdot \mathbf{A}'_{\Delta x} \right] d\xi. \, dx dt$$

$$= I_{\Delta S} - \sum_{ij} \int \int_{\mathcal{R}_{ij}} \varphi_t \int_{t_j}^t \left[g(\mathbf{A}_{ij}, \hat{u}(\xi - t_j, u_{\Delta x}^{RP}(x,\xi)), x) \right.$$

$$\left. - \frac{\partial f}{\partial \mathbf{A}}(\mathbf{A}_{ij}, \hat{u}(\xi - t_j, u_{\Delta x}^{RP}(x,\xi))) \cdot \mathbf{A}'_{\Delta x} \right] d\xi. \, dx dt$$

$$= O(\Delta x^2) - \sum_{ij} \int_{\mathcal{R}_i} \left\{ \varphi(x, t_{j+1}) \int_{t_j}^{t_{j+1}} \left[g(\mathbf{A}_{ij}, \hat{u}(\xi - t_j, u_{\Delta x}^{RP}(x,\xi)), x) \right. \right.$$

$$\left. - \frac{\partial f}{\partial \mathbf{A}}(\mathbf{A}_{ij}, \hat{u}(\xi - t_j, u_{\Delta x}^{RP}(x,\xi))) \cdot \mathbf{A}'_{\Delta x} \right] d\xi$$

$$\left. - \int_{t_j}^{t_{j+1}} \varphi \left[g(\mathbf{A}_{ij}, u_{\Delta x}, x) - \frac{\partial f}{\partial \mathbf{A}}(\mathbf{A}_{ij}, u_{\Delta x}) \cdot \mathbf{A}'_{\Delta x} \right] dt \right\} dx,$$

$$= O(\Delta x^2) - \sum_{ij} \int_{\mathcal{R}_i} \left\{ \varphi(x, t_{j+1}) \int_{t_j}^{t_{j+1}} \left[g(\mathbf{A}_{ij}, \hat{u}(\xi - t_j, u_{\Delta x}^{RP}(x, t_{j+1})), x) \right. \right.$$

$$\left. - \frac{\partial f}{\partial \mathbf{A}}(\mathbf{A}_{ij}, \hat{u}(\xi - t_j, u_{\Delta x}^{RP}(x, t_{j+1}))) \cdot \mathbf{A}'_{\Delta x} \right] d\xi \right\} dx$$

$$+ I_4 + I_5 \tag{4.9.15}$$

where

$$I_4 = \sum_{ij} \int_{\mathcal{R}_i} \left\{ \varphi(x, t_{j+1}) \int_{t_j}^{t_{j+1}} \left[g(\mathbf{A}_{ij}, \hat{u}(\xi - t_j, u_{\Delta x}^{RP}(x, t_{j+1})), x) \right. \right.$$

$$- g(\mathbf{A}_{ij}, \hat{u}(\xi - t_j, u_{\Delta x}^{RP}(x,\xi)), x)$$

$$- \frac{\partial f}{\partial \mathbf{A}}(\mathbf{A}_{ij}, \hat{u}(\xi - t_j, u_{\Delta x}^{RP}(x, t_{j+1}))) \cdot \mathbf{A}'_{\Delta x}$$

$$\left. \left. + \frac{\partial f}{\partial \mathbf{A}}(\mathbf{A}_{ij}, \hat{u}(\xi - t_j, u_{\Delta x}^{RP}(x,\xi))) \cdot \mathbf{A}'_{\Delta x} \right] d\xi \right\} dx,$$

and

$$I_5 = \sum_{ij} \int\int_{\mathcal{R}_{ij}} \varphi \left[g(\mathbf{A}_{ij}, u_{\Delta x}, x) - \frac{\partial f}{\partial \mathbf{A}}(\mathbf{A}_{ij}, u_{\Delta x}) \cdot \mathbf{A}'_{\Delta x} \right] dx dt. \quad (4.9.16)$$

Note that

$$|I_4| \le |\varphi|_\infty \sum_{ijl} C |\gamma^l_{ij}| \Delta x \Delta t \le |\varphi|_\infty C \Delta x \Delta t \sum_j V$$

$$= |\varphi|_\infty C \Delta x \Delta t V \frac{T}{\Delta t} = O(\Delta x),$$

where the sum on j is taken over t_j in $Supp(\varphi)$. Substituting (4.9.13) and (4.9.15) into (4.9.12), we have

$$\varepsilon(u_{\Delta x}, \mathbf{A}_{\Delta x}, \varphi) = O(\Delta x) + \varepsilon_1(u_{\Delta x}, \mathbf{A}_{\Delta x}, \varphi)$$

$$- \sum_{ij} \int\int_{\mathcal{R}_{ij}} \varphi \frac{\partial f}{\partial \mathbf{A}}(\mathbf{A}_{ij}, u_{\Delta x}) \cdot \mathbf{A}'_{\Delta x} dx dt \quad (4.9.17)$$

$$+ \sum_{ij} \int_{\mathcal{R}_j} \varphi(x_{i+\frac{1}{2}}, t) \left\{ f(\mathbf{A}_{i+1,j}, u^{RP}_{\Delta x}(x_{i+\frac{1}{2}}, t)) - f(\mathbf{A}_{ij}, u^{RP}_{\Delta x}(x_{i+\frac{1}{2}}, t)) \right\} dt.$$

It is evident now that

$$\varepsilon(u_{\Delta x}, \mathbf{A}_{\Delta x}, \varphi) = \varepsilon_1(u_{\Delta x}, \mathbf{A}_{\Delta x}, \varphi) + O(\Delta x). \quad (4.9.18)$$

We use Glimm's technique to show that $\varepsilon_1(u_{\Delta x}, \mathbf{A}_{\Delta x}, \varphi) = O(\Delta x)$, c.f. [10].

To estimate ε_1, write $\varepsilon \equiv \varepsilon_1(\Delta x, \varphi, \mathbf{a})$ to display its dependence on $(\Delta x, \phi, \mathbf{a})$, where $\mathbf{a} \in \mathbf{\Pi}$ is the sample sequence, c.f. (4.3.24) above. Set

$$\varepsilon^j_1(\Delta x, \varphi, \mathbf{a}) = \int_{r_0}^\infty \varphi(x, t_j) \left\{ u_{\Delta x}(x, t_j^+) - u_{\Delta x}(x, t_j^-) \right\} dx. \quad (4.9.19)$$

Now since $u(x, t) = \Psi^{-1} \cdot \mathbf{\Phi} \cdot \mathbf{z}(x, t)$, it follows from (4.8.70) that there exists a constant V such that $TV_{[x, x+L]} u_{\Delta x}(\cdot, t) \le V$ on $r_0 \le x < \infty$, $t < T$. Using this, the following lemma gives estimates for ε_1 and ε^j_1.

Lemma 13 Let $\mathbf{a} \in \mathbf{\Pi}$, and let $\varphi \in C_0 \cap L^\infty$ be a test function in the space of continuous functions of compact support in $r_0 \le x < \infty$, $0 \le t < T$. Suppose $TV_{[x, x+L]} u_{\Delta x}(\cdot, t) \le V$ for all $x \ge r_0$, $t < T$. Then

$$|\varepsilon^j_1(\Delta x, \varphi, \mathbf{a})| \le V \frac{diam\,(spt\varphi)}{L} \Delta x \|\varphi\|_\infty, \quad (4.9.20)$$

and

$$|\varepsilon_1(\Delta x, \varphi, \mathbf{a})| \leq \frac{V\Lambda}{L} \left(diam\,(spt\varphi)\right)^2 \|\varphi\|_\infty. \qquad (4.9.21)$$

Proof: Since $[u_{\Delta x}](x, t_j)$ is bounded by the sum of the wave strengths from $x_{i-\frac{1}{2}}$ to $x_{i+\frac{1}{2}}$ for each x at time $t = t_j$, it follows that

$$|\varepsilon_1^j| \leq \|\varphi\|_\infty \sum_{i,p} \|\gamma_{ij}^p\|_u \Delta x \leq \|\varphi\|_\infty V \frac{diam\,(spt\varphi)}{L} \Delta x, \qquad (4.9.22)$$

where $\|\gamma\|_u$ denotes the strength of a wave in u-space. This verifies (4.9.20). Consequently, if J is the smallest j so that $t = t_j$ upper bounds the support of φ, then $J = T/\Delta t$, where $T = J\Delta t$, and

$$|\varepsilon_1| \leq \sum_{j=1}^{J} |\varepsilon_1^j| \leq \frac{T}{\Delta t} \|\varphi\|_\infty \Delta x V \frac{diam\,(spt\varphi)}{L}$$

$$\leq \frac{V\Lambda}{L} \left(diam\,(spt\varphi)\right)^2 \|\varphi\|_\infty$$

where $\Delta x/\Delta t \leq \Lambda$. □

We next show that ε_1^j, when taken as a function of a_j, has mean zero.

Lemma 14 *For approximate solutions $u_{\Delta x}$,*

$$\int_0^1 \varepsilon_1^j \, da_j = 0 \qquad (4.9.23)$$

Proof: The proof follows from Fubini's theorem.

$$\int_0^1 \varepsilon_1^j \, da_j = \int_0^1 \sum_0^\infty \int_{x_{i-\frac{1}{2}}}^{x_{i+\frac{1}{2}}} [u_{\Delta x, \mathbf{a}}(x_i + a_j \Delta x, t_j) - u_{\Delta x, \mathbf{a}}(x, t_j)] \, dx da_j$$

$$= \sum_{i=0}^\infty \left\{ \int_{x_{i-\frac{1}{2}}}^{x_{i+\frac{1}{2}}} \int_0^1 u_{\Delta x, \mathbf{a}}(x_i + a_j \Delta x, t_j) \, da_j dx \right.$$

$$\left. - \int_0^1 \int_{x_{i-\frac{1}{2}}}^{x_{i+\frac{1}{2}}} u_{\Delta x, \mathbf{a}}(x, t_j) \, dx da_j \right\}$$

$$= 0,$$

which was to be proved. (Here we used $u_{\Delta x, \mathbf{a}}$ to express the dependence of the approximate solution $u_{\Delta x}$ on the sample sequence $\mathbf{a}$.) □

We now show that the functions ε_1^j are orthogonal, when taken as elements of $L^2(\mathbf{\Pi})$.

Lemma 15 *Suppose φ has compact support, and is piecewise constant on rectangles $\mathcal{R}_{ij}$. Then if $j_1 \neq j_2$, we have $\varepsilon_1^{j_1} \perp \varepsilon_1^{j_2}$ where orthogonality is with respect to the inner product on $L^2(\mathbf{\Pi})$.*

Proof: Using Lemma 15 in calculating the inner product

$$
\langle \varepsilon_1^{j_1}, \varepsilon_1^{j_2} \rangle = \int \varepsilon_1^{j_1} \varepsilon_1^{j_2} \left(\Pi \, da_j \right) = \int \left(\int \varepsilon_1^{j_1} \varepsilon_1^{j_2} \, da_{j_2} \right) \Pi_{j \neq j_2} \, da_j
$$

$$
= \int \varepsilon_1^{j_1} \left(\int \varepsilon_1^{j_2} \, da_{j_2} \right) \Pi_{j \neq j_2} \, da_j
$$

$$
= 0,
$$

verifying orthogonality. $\square$

It follows immediately from Lemma 15 that

$$
\|\varepsilon_1\|_2^2 = \sum_j \|\varepsilon_1^j\|_2^2, \tag{4.9.24}
$$

which we use in our next theorem to finally show that there is a subsequence so that $\varepsilon_1 \to 0$ as $\Delta x \to 0$ for almost any $\mathbf{a} \in \mathbf{\Pi}$.

Theorem 16 *Suppose that $TV_{[x, x+L]} u_{\Delta x}(\cdot, t) \leq V$ for all $r_0 \leq x < \infty$, $0 \leq t < T$. Then there is a null set $N \subset \mathbf{\Pi}$ and a sequence Δx_k such that for all $\mathbf{a} \in \mathbf{\Pi} - N$ and $\varphi \in C_0^1(t > 0)$, we have $\varepsilon_1(\Delta x, \varphi, \mathbf{a}) \to 0$ as $k \to \infty$.*

Proof: Combining (4.9.24) and (4.9.20), and using the fact that $\int_{\mathbf{\Pi}} da = 1$, we have

$$
\|\varepsilon_1(\Delta x, \varphi, \mathbf{a})\|_2^2 = \sum_j \|\varepsilon_1^j(\Delta x, \varphi, \mathbf{a})\|_2^2
$$

$$
\leq \sum_j \|\varepsilon_1^j(\Delta x, \varphi, \mathbf{a})\|_\infty^2
$$

$$
\leq \sum_{j=0}^J V^2 (\Delta t_k)^2 \|\varphi\|_\infty^2 \frac{(diam\,(spt\varphi))^2}{L^2}
$$

$$
\leq V^2 (\Delta t_k) \frac{(diam\,(spt\varphi))^3}{L^2} \|\varphi\|_\infty^2,
$$

and hence, for piecewise constant φ with compact support, there is a sequence $\Delta x_k \to 0$ such that $\varepsilon_1 \to 0$ in L^2. If φ is continuous with compact support, then by (4.9.21),

$$
\|\varepsilon_1\|_2 \leq \|\varepsilon_1\|_\infty \leq C\|\varphi\|_\infty. \tag{4.9.25}
$$

Let $\{\varphi_l\}$ be a sequence of piecewise constant functions with constant support whose closure relative to the infinity norm contains the space of test functions that are continuous with compact support. For each l, there is a null set $N_l \subset \mathbf{\Pi}$ and a sequence $\Delta x_{k_n(l)} \to 0$ such that $\varepsilon_1 \to 0$ pointwise, for all $a \in \mathbf{\Pi} - N_l$. Set $N = \bigcup_l N_l$, and let $a \in \mathbf{\Pi} - N$. By a diagonalization process, we can find a subsequence, Δx_k, such that for each l, $\varepsilon_1 \to 0$ as $k \to \infty$. If ψ is any test function, then if $a \in \mathbf{\Pi} - N$, we have

$$|\varepsilon_1(\Delta x, \psi, \mathbf{a})| \leq |\varepsilon_1(\Delta x, \psi - \varphi_l, \mathbf{a})| + |\varepsilon_1(\Delta x, \varphi_l, \mathbf{a})|$$
$$\leq Const.\|\psi - \varphi_l\|_\infty + |\varepsilon_1(\Delta x, \varphi_l, \mathbf{a})|.$$

It is now clear that given $\varepsilon > 0$, there exists $N \in \mathbf{N}$ so that if $i, l \geq N$, then $|\varepsilon_1(\Delta x, \psi, \mathbf{a})| \leq \varepsilon$. □

4.10 Concluding Remarks

The shock wave solutions of the Einstein equations constructed in Chapter 4 have the property that the components of the gravitational metric tensor are only Lipschitz continuous, that is, in the class $C^{0,1}$, (functions whose 0-order derivatives are continuous with Holder exponent 1, [9]), at shock waves. Now this is one derivative *less* smooth than the Einstein equations suggest it should be, and in fact, the singularity theorems in [14] presume that metrics are in the smoothness class $C^{1,1}$, one degree *smoother* than the solutions we have constructed, c.f. [14], page 284. This suggests the following open mathematical question, [13]: *Given a weak solution of the Einstein equations for which the metric components are only Lipschitz continuous functions of the coordinates, under what conditions does there exist a coordinate transformation that improves the regularity of the metric components from $C^{0,1}$ to $C^{1,1}$?* For a single, non-lightlike shock surface, it is known that such a coordinate transformation always exists, and the transformed coordinates can be taken to be Gaussian normal coordinates at the shock, [28, 31]. However, the solutions we construct in Chapter 4 allow for arbitrary numbers of interacting shock waves of arbitrary strength, and at points of interaction, the Gaussian normal coordinate systems break down. Thus, in particular, we ask whether the solutions that we construct in Chapter 4 can be improved to $C^{1,1}$ by coordinate transformation? If such a coordinate transformation does *not* exist, then solutions of the Einstein equations are one degree less smooth than expected. If such a transformation does exist, then it defines a mapping that takes weak (shock wave) solutions of the Einstein equations to strong solutions.

This question goes to the heart of the issue of the regularity of solutions of the Einstein equations. Indeed, the Einstein equations are inherently hyperbolic in character because of finite speed of propagation; i.e., no information can propagate faster than the speed of light. It follows that, unlike Navier-Stokes type parabolic regularizations of the classical compressible Euler equations, augmenting the perfect fluid assumption by incorporating the effects of viscosity and dissipation into Einstein's theory of gravity, cannot alter the fundamental hyperbolic character, (finite speed of propagation), of the Einstein equations. Thus, even when dissipative effects are accounted for, it is not clear apriori that solutions of the Einstein equations will in general be more regular than the solutions that we have constructed here.

In summary, if a transformation exists that impoves the regularity of shock wave solutions of the Einstein equations from the class $C^{0,1}$ up to the class $C^{1,1}$, then it defines a mapping that takes weak solutions of the Einstein equations to strong solutions. It then follows that in general relativity, the theory of distributions and the Rankine Hugoniot jump conditions for shock waves need not be imposed on the compressible Euler equations as extra conditions on solutions, but rather must follow as a logical consequence of the strong formulation of the Einstein equations by themselves. If such a transformation does not always exist, then solutions of the Einstein equations are one degree less regular than previously assumed.

References

1. A.M. Anile, *Relativistic Fluids and Magneto-Fluids*, Cambridge Monographs on Mathematical Physics, Cambridge University Press, 1989.
2. D. Chae, *Global existence of spherically symmetric solutions to the coupled Einstein and nonlinear Klein-Gordon system*, Class. Quantum Grav., **18**(2001), pp. 4589-4605.
3. Y. Choquet-Bruhat and R. Geroch, *Global aspects of the Cauchy problem in general relativity*, Commun. Math. Phys., **14**(1969), pp. 329-335.
4. Y. Choquet-Bruhat and J. York, *The Cauchy problem*, in *General Relativity and gravitation*, edited by A. Held, Volume 1, pp. 99-172, Plenum Pub., New York, 1980.
5. D. Christodoulou, *Global existence of generalized solutions of the spherically symmetric Einstein-scalar equations in the large*, Commun. Math. Phys., **106**(1986), pp. 587-621.
6. D. Christodoulou and S. Klainerman, *The global nonlinear stability of the Minkowski spacetime*, Princeton University Press, Princeton, 1993.
7. R. Courant and K. Friedrichs, *Supersonic Flow and Shock–Waves*, Wiley-Interscience, 1948. 1972.
8. B.A. Dubrovin, A.T. Fomenko and S.P.Novikov, *Modern Geometry-Methods and Applications*, Springer Verlag, 1984.
9. Lawrence C. Evans, *Partial Differential Equations, Vol 3A*, Berkeley Mathematics Lecture Notes, 1994.
10. J. Glimm, *Solutions in the large for nonlinear hyperbolic systems of equations*, Comm. Pure Appl. Math., **18**(1965), pp. 697-715.
11. J. Groah, *Solutions of the Relativistic Euler equations in non-flat spacetimes*, Doctoral Thesis, UC-Davis.
12. J. Groah and B. Temple, *A shock wave formulation of the Einstein equations*, Methods and Applications of Analysis, **7**, No. 4, (2000).
13. J. Groah and B. Temple, *shock wave solutions of the spherically symmetric Einstein equations with perfect fluid sources: existence and consistency by a locally inertial Glimm scheme*, Memoirs of the AMS, **172**, No. 813, pp. 1-84.
14. S.W. Hawking and G.F.R. Ellis, *The Large Scale Structure of Spacetime*, Cambridge University Press, 1973.
15. W. Israel, *Singular hypersurfaces and thin shells in General Relativity*, IL Nuovo Cimento, Vol. XLIV B, N. 1, 1966, 1-14.

16. S. Kind and J. Ehlers, *Initial boundary value problem for spherically symmetric Einstein equations for a perfect fluid*, Class. Quantum Grav., **18**(1957), pp. 2123–2136.

17. P.D. Lax, *Hyperbolic systems of conservation laws, II*, Comm. Pure Appl. Math., **10**(1957), pp. 537–566.

18. P.D. Lax, *Shock–waves and entropy*. In: Contributions to Nonlinear Functional Analysis, ed. by E. Zarantonello, Academic Press, 1971, pp. 603-634.

19. M. Luskin and B. Temple, *The existence of a global weak solution of the water-hammer problem*, Comm. Pure Appl. Math. Vol. 35, 1982, pp. 697-735.

20. T. Makino, K. Mizohata and S. Ukai, *Global weak solutions of the relativistic Euler equations with spherical symmetry*, Japan J. Ind. and Appl. Math., Vol. 14, No. 1, 125-157 (1997).

21. C. Misner, K. Thorne, and J. Wheeler, *Gravitation*, Freeman, 1973.

22. T. Nishida, *Global solution for an initial boundary value problem of a quasilinear hyperbolic system*, Proc. Jap. Acad., **44**(1968), pp. 642-646.

23. T. Nishida and J. Smoller *Solutions in the large for some nonlinear hyperbolic conservation laws*, Comm. Pure Appl. Math., **26**(1973), pp. 183-200.

24. J.R. Oppenheimer and G.M. Volkoff, *On massive neutron cores*, Phys. Rev., **55**(1939), pp. 374-381.

25. *Theorems on existence and global dynamics for the Einstein equations*, A Rendall, Living Rev. Rel. **5** (2002) (update) and arXiv: gr-qc/0505133, Vo. 1 26 May, 2005.

26. J. Smoller, *Shock Waves and Reaction Diffusion Equations*, Springer Verlag, 1983.

27. J. Smoller and B. Temple *Global solutions of the relativistic Euler equations*, Comm. Math. Phys., **157**(1993), p. 67-99.

28. J. Smoller and B. Temple, *Shock–wave solutions of the Einstein equations: the Oppenheimer-Snyder model of gravitational collapse extended to the case of nonzero pressure*, Arch. Rat. Mech. Anal., **128** (1994), pp 249-297.

29. J. Smoller and B. Temple, *Astrophysical shock–wave solutions of the Einstein equations*, Phys. Rev. D, **51**, No. 6 (1995).

30. *shock wave solutions of the Einstein equations: A general theory with examples*, with J. Smoller, Proceedings of European Union Research Network's 3rd Annual Summerschool, Lambrecht (Pfalz) Germany, May 16-22, 1999.

31. *Solving the Einstein equations by Lipschitz continuous metrics*, with J. Smoller, Handbook of Mathematical Fluid Dynamics Vol. 2, Edited by S. Friedlander and D. Serre, pp. 501-597, North-Holland.

32. A. Taub, *Approximate Solutions of the Einstein equations for isentropic motions of plane-symmetric distributions of perfect fluids,*, Phys. Rev., **107, no. 3**, (1957), pp. 884-900.

33. R. Tolman, *Relativity, Thermodynamics and Cosmology*, Oxford University Press, 1934.

34. R.M. Wald, *General Relativity*, University of Chicago Press, 1984.

35. S. Weinberg, *Gravitation and Cosmology: Principles and Applications of the General Theory of Relativity*, John Wiley & Sons, New York, 1972.

Index